Peter A. Hall, David A. Levison and
Nicholas A. Wright (Eds)

Assessment of Cell Proliferation in Clinical Practice

With 36 Figures

Springer-Verlag
London Berlin Heidelberg New York
Paris Tokyo Hong Kong
Barcelona Budapest

Peter A. Hall, BSc, MD, MRCPath
Professor of Histopathology, Department of Histopathology,
UMDS, St Thomas' Hospital, London SE1 7EH

David A. Levison, MD, FRCPath
Professor of Histopathology, Department of Histopathology,
UMDS, Guy's Hospital, London SE1 9RT

Nicholas A. Wright, MA, MD, PhD, DSc, FRCPath
Director of Clinical Research, Imperial Cancer Research Fund,
Lincoln's Inn Fields, London WC2A 3PN, and Professor and
Director, Department of Histopathology, Royal Postgraduate
Medical School, Hammersmith Hospital, London W12 0NN

ISBN-13: 978-1-4471-3192-2

British Library Cataloguing in Publication Data
Assessment of cell proliferation in clinical
  practice.
  I. Hall, Peter A., *1958–*
  II. Levison, David A., *1944–*
  III. Wright, Nicholas A., *1943–*
  574.8762
ISBN-13: 978-1-4471-3192-2

Library of Congress Cataloging-in-Publication Data
Assessment of cell proliferation in clinical practice/ edited by
  Peter A. Hall, David A. Levison, Nicholas A. Wright.
      p.   cm.
  Includes index.
  ISBN-13: 978-1-4471-3192-2   e-ISBN-13: 978-1-4471-3190-8
  DOI: 10.1007/978-1-4471-3190-8
  1. Cancer cells—Proliferation—Measurement. 2. Cancer—
Prognosis. I. Hall, Peter A. (Peter Anthony), 1958– .
II. Levison, David A. (David Annan), 1944– . III. Wright,
Nicholas A.
  [DNLM: 1. Cell Differentiation. 2. Cell Division. 3. Cell
Transformation, Neoplastic. QZ 202 A845]
  RC269.7.A87 1991
  616.99'4071—dc20
  DNLM/DLC
  for Library of Congress   91-4875
                                                       CIP

Typeset by Best-set Typesetter Ltd, Hong Kong

28/3830-543210 Printed on acid-free paper

This book is dedicated to the memory of
Dr Nicholas McNally
(1939 to 1991)

Preface

For more than three decades the methods for assessing cell proliferation have been largely the preserve of experimental biologists, and in their hands such techniques have contributed greatly to our understanding of the dynamic organisation of normal and pathological tissues. In recent years, with the advent of novel methodologies, there has been increased interest among both pathologists and clinicians, particularly oncologists and others interested in neoplasia, in assessing cell proliferation. This interest has been stimulated by the possibility that indices of cell proliferation *may* have direct clinical relevance, for example in being useful predictors of outcome in patients with certain forms of malignancy. In addition, interest in assessing cell proliferation has been fuelled by the tremendous advances in our understanding of the mechanisms of cell proliferation and their deregulation in pathological processes. Consequently, the time is ripe for a monograph critically reviewing the available methods for assessing cell proliferation, their potential and their problems.

We have been particularly concerned to present a balanced view of the advantages and disadvantages of different methods currently available for assessing cell proliferation. The assessment of cell proliferation often requires some familiarity with mathematical methods, but in this book we have attempted to keep detailed mathematical analyses to a minimum. We have asked exponents of each of the well-recognised methods to critically review the techniques and the ways in which they may be applied to clinical material. Assessing cell proliferation cannot be performed in isolation; therefore the book begins with three chapters designed to put the assessment of cell proliferation into the context of contemporary biology. First, Dr Robert Brooks has comprehensively reviewed our current knowledge of the molecular basis of the cell cycle and its control. Drs Nick Lemoine and Donal Hollywood have discussed the range of alterations that may occur in the regulation of growth control in neoplasia and Dr Ansari and Professor Hall have focussed upon the general kinetic and spatial architecture of tissues, and in particular the biology of stem cells.

Dr Robin Dover has presented an overview of thymidine labelling methods, surely still the "gold standard" for kinetic assays, together with a brief overview of other methods. Each of these methods is then discussed in detail. Dr Cecily Quinn and Professor Nick Wright deal with mitosis counting, Drs Richard Camplejohn and James Macartney with flow cytometry and Dr George Wilson and the late Dr Nicholas McNally with the application of bromodeoxyuridine labelling in vivo. This latter chapter provides a detailed account of exciting new methods that allow the generation of data relating to *rates of cellular proliferation* rather than the simple descriptive state parameters. Drs Yu, Woods and Professor Levison have focussed on immunohistological methods for demonstrating proliferating cells, and then Professor Underwood critically assesses the possible value of analysing nucleolar organiser regions.

The final two chapters of the book are designed to give an indication of the possible value of the information gained in the context of clinical practice (Drs Susan O'Reilly and Michael Richards) and in the understanding of the mode of action of anticancer therapy (Dr Walter Gregory).

We hope that this book will be of principal interest to all those who are contemplating employing methods for assessing cell proliferation in clinical material or in other in vivo situations, whether they be pathologists, oncologists, surgeons or physicians. The book should also be of interest to other biologists in a wide range of fields who may need to have recourse to assessing cell proliferation.

April 1991   Peter A. Hall
                                                        David A. Levison
                                                        Nicholas A. Wright

Contents

Contributors

B. Ansari, MD
Clinical Research Fellow, Department of Histopathology, Royal
Postgraduate Medical School, Hammersmith Hospital, Du Cane
Road, London W12 0NN

R.F. Brooks, BSc, PhD
Lecturer, Anatomy and Human Biology, Division of Biomedical
Sciences, King's College London, Strand, London WC2R 2LS

R.S. Camplejohn, MSc, PhD
Senior Lecturer, Richard Dimbleby Department of Cancer Research,
UMDS, St Thomas' Hospital, London SE1 7EH

R. Dover, BSc, PhD
Postdoctoral Research Fellow, ICRF Histopathology Unit, Imperial
Cancer Research Fund, Lincoln's Inn Fields, London WC2A 3PN

W.M. Gregory, BSc, PhD
Senior Lecturer, ICRF Clinical Oncology Unit, UMDS, Guy's
Hospital, London SE1 9RT

P.A. Hall, BSc, MD, MRCPath
Professor of Histopathology, Department of Histopathology,
UMDS, St Thomas' Hospital, Lambeth Palace Road, London
SE1 7EH

D.P. Hollywood, MB BCh, BAO, MRCPI, FFR(RCSI), FRCR
Clinical Research Fellow, ICRF Molecular Oncology Group,
Hammersmith Hospital, Du Cane Road, London W12 0NN

N.R. Lemoine, BSc, MBBS, PhD
Clinical Scientist, ICRF Molecular Oncology Group, Hammersmith
Hospital, Du Cane Road, London W12 0NN

D.A. Levison, MD, FRCPath
Professor and Head, Department of Histopathology, UMDS, Guy's
Hospital, London SE1 9RT

J.C. Macartney, MD, FRCPath
Consultant, Department of Pathology, The Alexandra Hospital,
Redditch B98 7UB

N.J. McNally, BSc, PhD
Formerly Senior Scientist, CRC Gray Laboratory, PO Box 100,
Mount Vernon Hospital, Northwood, Middlesex HA6 2JR

S.M. O'Reilly, MB ChB, MRCP
Lecturer and Honorary Senior Registrar, Department of Medical
Oncology, Charing Cross Hospital, Fulham Palace Road, London
W6 8RF

C.M. Quinn, MB ChB, BAO, MRCP, MRCPath
Senior Registrar, Department of Histopathology, Royal Post-
graduate Medical School, Hammersmith Hospital, Du Cane Road,
London W12 0NN

M.A. Richards, MA, MD, MRCP
Senior Lecturer, ICRF Clinical Oncology Unit, UMDS, Guy's
Hospital, London SE1 9RT

J.C.E. Underwood, MD, FRCPath
Professor and Director, Department of Pathology, University of
Sheffield Medical School, Beach Hill Road, Sheffield S10 2RX

G.D. Wilson, BSc, PhD
Senior Scientist, CRC Gray Laboratory, PO Box 100, Mount
Vernon Hospital, Northwood, Middlesex HA6 2JR

A.L. Woods, MB BS, MRCP
Registrar, Department of Histopathology, UMDS, Guy's Hospital,
London SE1 9RT

N.A. Wright, MA, MD, PhD, DSc, FRCPath
Director of Clinical Research, Imperial Cancer Research Fund,
Lincoln's Inn Fields, London WC2A 3PN, and Professor and
Director, Department of Histopathology, Royal Postgraduate
Medical School, Hammersmith Hospital, Du Cane Road, London
W12 0NN

C.C.-W. Yu, MA, MBChB, MRCP
Lecturer, Department of Histopathology, UMDS, Guy's Hospital,
London SE1 9RT

Abbreviations and Conventions

AgNOR	argyrophilic nucleolar organiser region
BUdR*	5-bromo-2-deoxyuridine
C	haploid DNA content
$CDC^\dagger$	cell division cycle (*Saccharomyces cerevisiae*)
$cdc^\dagger$	cell division cycle (*Schizosaccharomyces pombe*)
CHART	continuous hyperfractionated accelerated radiotherapy
CV	coefficient of variation
DAPI	4,6-diamidino-2-phenylindole-dihydrochloride
ϕ	cell loss factor
FCM	flow cytometry
FLM (or PLM)	fraction of labelled mitosis
G_0	post-mitotic proliferative. quiescence
G_1	first gap period preceding S phase
G_2	second gap, following S phase and preceding mitosis
GF	growth fraction
HPF	high-powered field
I_p	index of proliferation (= GF)
IUdR	5-iodo-2-deoxyuridine
K_B	birth rate of cell population
kDa	kilodaltons
K_G	growth rate of cell population
K_L	cell loss rate
λ	correction factor for age distribution of cell population
ln	$\log_e$
LI	labelling index (the fraction of cycling cells)

---

* A range of abbreviations is employed and no universal consensus has emerged. We have adopted a uniform style in this book based on that employed in the Gray Laboratory

† By convention these abbreviations relate to the genes themselves. With the cloning of genes in other eukaryotes, including man, the terminology is becoming confused. The confusion is compounded by the use of the abbreviation *Cdc* to represent the protein product of a *cdc* gene!

MI	mitotic index
MPF	mitosis-promoting factor (sometimes maturation-promoting factor)
mRNA	messenger RNA
NOR	nucleolar organiser region
NORAP	nucleolar organiser region associated protein
P cells	cells born into the proliferative compartment
PCNA[‡]	proliferating cell nuclear antigen
PI	propidium iodide
Q cells	those born into the non-proliferative compartment
RM	relative movement
r_M	rate of entry into mitosis
rRNA	ribosomal RNA
r_S	rate of entry into S phase
SPF	S-phase fraction
T_c	cell cycle time
T_d	population doubling time
TLI	thymidine labelling index
T_m	duration of mitosis
T_{pot}	potential doubling time
T_S	duration of S phase
T_t	turnover time

[‡]The term cyclin has been used to denote PCNA. However, there is now general agreement that the term cyclin should be reserved for a family of regulatory proteins which accumulate during interphase only to be destroyed during mitosis (see Chap. 1 and Brooks et al., 1989, J Cell Sci Suppl 12: ii).

1 Regulation of the Eukaryotic Cell Cycle

R.F. BROOKS

Introduction

Prior to the 1950s, the first indication that a cell was committed to divide was the onset of mitosis itself, marked by the dramatic events of chromosome condensation, nuclear envelope breakdown and spindle formation. With the advent of suitable radioactive tracers, it soon became clear that events relevant to cell division, notably DNA replication, actually began many hours before mitosis (Howard and Pelc 1953). This led to the subdivision of interphase into the now familiar phases of the cell cycle, namely: G_1, the gap between mitosis and the start of DNA replication; S phase, the period of DNA synthesis; and G_2, the gap between the completion of DNA synthesis and the onset of mitosis, M. Following measurements of the duration of each phase, it quickly became apparent that for many eukaryotic cell types (and certainly, for the somatic cells of vertebrates), regulation of the cycle occurs primarily in G_1 (Smith 1982). Thus, though not invariant, the duration of S + G_2 + M generally changes far less than does the duration of G_1 with changes in proliferation rate. Furthermore, cells which cease proliferation, either reversibly (e.g. hepatocytes, lymphocytes) or permanently (e.g. neurones), come to arrest in G_1. Passage through G_1 requires an adequate nutritional environment and, in the case of vertebrate somatic cells, stimulation by an appropriate set of polypeptide growth factors. However, once past a certain point (referred to variously as the restriction point, or START), the cell is committed to divide and removal of nutrients or growth factors fails to prevent entry into S phase (Hartwell 1974; Pardee 1974).

With the realisation that the main control point of the cell cycle lay in G_1, attention shifted away from mitosis to the environmental cues regulating proliferation. Motivated by the desire to understand cancer, work with animal cells has predominated, and during the past two decades outstanding advances have been made in identifying growth factors and the pathways by which they act to control growth (see Chap. 2). These pathways are believed to involve the products of proto-oncogenes – genes whose deranged expression leads to unrestrained cell proliferation. Nevertheless, despite intensive study, the point of convergence of these pathways remains elusive, in part because there has been no biochemical assay for the passage of START. Recently, however, there have been signs of progress. Ironically, this has come not from the extensive work on growth factors and oncogenes but from the hitherto less fashionable studies of the regulation of mitosis. During the past 3 years, the broad outlines of the mechanism controlling the G_2/M transition have emerged (Dorée 1990; Hunt 1989a; Lewin 1990; Nurse 1990; Pines and Hunter 1990a). It is now clear that entry into mitosis depends on the activation of a specific protein kinase whose key components, $p34^{cdc2}$ and cyclins, have been highly conserved in all eukaryotes so far examined. Surprisingly, genetic studies in yeasts have shown that $p34^{cdc2}$ is also required at START in addition to mitosis. This conclusion has been reinforced by more recent biochemical evidence from vertebrate systems, and it seems increasingly likely that there will be many parallels between the mechanisms controlling mitosis and START. In this chapter, the role of $p34^{cdc2}$ in the regulation of mitosis will therefore be considered in some detail before returning to the question of what commits a cell to begin the cell cycle.

Mitosis-Promoting Factor

Frog oocytes are arrested at prophase of the first meiotic division and normally mature to eggs (arrested at metaphase of the second meiotic division) in response to progesterone released by the surrounding follicle cells (Kirschner et al. 1985). Progesterone-induced maturation takes about 6 h and is blocked by cycloheximide, indicating a need for new protein synthesis. Much of our current understanding of mitotic control mechanisms stems from the discovery by Masui and Markert (1971) and Smith and Ecker (1971) that oocytes could be made to mature in the absence of progesterone, and in less than 2 h, by injecting into them a small amount of cytoplasm from mature eggs. The factor responsible for this was termed maturation-promoting factor (MPF) and was evidently present at high levels in eggs since the cytoplasmic extracts could be diluted many-fold and still retain the capacity to induce oocyte maturation (Gerhart et al. 1984). Maturation induced by injection of a small amount of MPF also led to the development of the same high levels of MPF in recipient oocytes, and this amplification occurred even in the presence of cycloheximide (Gerhart et al. 1984). Evidently, MPF induced the autoactivation of an inactive precursor (termed pre-MPF) already present in the oocyte. It also followed that the protein synthesis requirement for progesterone-induced maturation was not to make MPF itself but an activator of pre-MPF.

On fertilization, the level of MPF in *Xenopus* eggs falls precipitously, only to rise again transiently just before each cleavage (Gerhart et al. 1984; Wasserman and Smith 1978). This shows that MPF is not confined to meiosis but is also a feature of mitotic cycles. Indeed, mitotic cells of many other species (from yeasts to humans) contain MPF activity, as defined by the ability to induce *Xenopus* oocyte maturation on microinjection (Kishimoto et al. 1982; Nelkin et al. 1980; Sunkara et al. 1979; Tachibana et al. 1987; Weintraub et al. 1982). MPF is therefore a universal mitotic inducer and the term is better now taken to stand for mitosis-promoting factor.

Following its disappearance on fertilisation (or parthenogenetic activation) the reappearance of MPF depends on protein synthesis, and in the presence of cycloheximide the egg arrests in G_2. Injection of a large dose of MPF into such cycloheximide-treated eggs overcomes the arrest, enabling them to enter a mitotic-like state marked by chromosome condensation and nuclear envelope disruption (Newport and Kirschner 1984). The injected MPF is unstable and decays spontaneously, and with this the chromosomes decondense once more. Most significantly, they also resume DNA synthesis. Thus, in cycloheximide-treated *Xenopus* eggs, MPF is all that is needed to drive the entire cell cycle. It follows that initiators of DNA synthesis must be present constitutively in the egg, and the provision of MPF is sufficient to overcome the block that normally prevents the re-replication of G_2 DNA (presumably by stimulating entry into mitosis).

The *cdc2* Gene*

Purification of MPF proved to be very difficult because of the laboriousness of the assay (microinjection into oocytes) and because of its instability. In retrospect, the latter is not surprising since we now know that its activity depends on a complex interplay of phosphorylation and dephosphorylation reactions. Eventually, following the development of a more convenient in vitro assay, involving cell-free extracts of *Xenopus* eggs that could recapitulate many of the events of mitosis, Lohka et al. (1988) succeeded in substantially purifying MPF. A consistent feature of the most highly purified preparations was the presence of a protein of approximately 32 kDa. In a remarkable example of scientific convergence, this was found to be the *Xenopus* homologue of a protein ($p34^{cdc2}$) first identified through cell cycle genetics in yeasts.

Following the pioneering work of Hartwell (1974) on the isolation of cell division cycle (*cdc*) mutants, many genes required for progress through the cell cycle had been identified, particularly in the budding yeast (*Saccharomyces cerevisiae*) and fission yeast (*Schizosaccharomyces pombe*) (Nurse et al. 1976; Pringle and Hartwell 1982). One of these, the *cdc2* gene of *S. pombe*, attracted particular attention because, uniquely, it was required at two points of the cycle, at both START and mitosis (Nurse and Bissett 1981). Moreover, certain alleles of *cdc2* led to the advancement of mitosis (i.e. by shortening G_2) such that cells divided at a smaller size than normal (the so-called 'wee' phenotype). This suggested that *cdc2* was not merely required for mitosis but played an important part in controlling its timing.

The *cdc2* gene of *S. pombe* encodes a 34 kDa protein kinase with a specificity for serine/threonine (Simanis and Nurse 1986). Unexpectedly, it was found to be highly homologous to the *CDC28* gene of *S. cerevisiae*, so much so that the two gene products could function interchangeably in the two organisms (Beach et al. 1982; Booher and Beach 1986). This was surprising since the *CDC28*

---

* Cell division cycle; this applies to cyclins, referred to first on p. 5.

gene was initially isolated solely as a START mutation, and it is only very recently that suggestions of a role at mitosis in *S. cerevisiae* (Piggott et al. 1982) have been confirmed (Reed and Wittenberg 1990). Even more surprising was the discovery that antibodies that recognised both the *cdc2* and *CDC28* gene products (presumably directed against the most highly conserved epitopes) also recognised a 34 kDa protein in human cells (Draetta et al. 1987). This suggested an extraordinary degree of conservation, which was dramatically confirmed when the human homologue was found to complement *cdc2* mutations in fission yeast (Lee and Nurse 1987).

Evidence that $p34^{cdc2}$ was in fact a component of MPF came with the demonstration that the 32 kDa protein found in the most highly purified preparations of MPF cross-reacted with an antibody to $p34^{cdc2}$ (Gautier et al. 1988). Additional evidence has been provided by other aspects of yeast cell cycle genetics. In fission yeast, $p34^{cdc2}$ interacts physically with $p13^{suc1}$, the 13 kDa product of the *suc1* gene, which was originally identified as a suppressor of certain *cdc2* mutations (Brizuela et al. 1987). It was found that $p13^{suc1}$ coupled to Sepharose could specifically deplete *Xenopus* egg extracts of both $p34^{cdc2}$ and MPF activity (Dunphy et al. 1988). Although we are still largely ignorant of the function of $p13^{suc1}$ in vivo (see later) $p13^{suc1}$/Sepharose has subsequently proved to be an extremely useful reagent for purifying MPF and for studying the biochemistry of $p34^{cdc2}$ (Labbé et al. 1989a; Pondaven et al. 1990).

Histone H1 Kinase

Many proteins show increased phosphorylation at mitosis including histone H1. This led to the identification of a major Ca^{2+}-, cyclic nucleotide- and diacylglycerol-independent histone H1 kinase, whose activity peaked sharply at mitosis (reviewed in Matthews and Huebner 1985; Wu et al. 1986). This enzyme, known variously as "growth-associated" or "M-phase-specific" H1 kinase, is found in a wide variety of eukaryotic cells. It is now generally accepted to be equivalent to MPF, or at least a principal manifestation of it.

The marked correspondence between the kinetics of appearance and disappearance of H1 kinase and MPF provided one of the first indications that the two activities might be closely related (Dabauvalle et al. 1988; Labbé et al. 1988a; Meijer et al. 1987; Picard et al. 1985, 1987). This possibility accorded with the early demonstration of a large increase in phosphorylation following the injection of MPF into *Xenopus* oocytes (Maller et al. 1977), and with the acceleration of entry into mitosis following the addition of partially purified H1 kinase to cultures of *Physarum* (Ingliss et al. 1976). The connection was firmly established when the M-phase-specific H1 kinase was shown to co-purify with $p34^{cdc2}$ for both starfish oocytes (Arion et al. 1988; Labbé et al. 1988b) and mammalian cells (Brizuela et al. 1989; Langan et al. 1989). Like MPF, H1 kinase activity also bound to $p13^{suc1}$/Sepharose (Arion et al. 1988; Brizuela et al. 1989). Furthermore, starfish H1 kinase co-fractionated with MPF activity throughout its purification (Labbé et al. 1989a, b). Conversely, highly purified *Xenopus* MPF was associated with kinase activity towards histone H1 (Lohka et al. 1988).

There seems little doubt that histone H1 kinase activity is a major bio-chemical manifestation of MPF, and that this kinase activity is directly respon-sible for triggering the onset of mitosis. For assay in vitro, histone H1 is a convenient substrate, and is almost certainly a physiological one (Langan et al. 1989), phosphorylation of which may be involved in chromosome condensa-tion. Other substrates include the nuclear lamins (Peter et al. 1990; Ward and Kirschner 1990) which are solubilised by phosphorylation just before nuclear envelope breakdown. However, though probably important targets, neither histone H1 nor lamins are likely to be key regulatory substrates, central to the initiation of mitosis, in so far as both are absent from yeasts in which $p34^{cdc2}$ was first identified (Arion et al. 1988). Instead, it seems probable that the targets of $p34^{cdc2}$ kinase (henceforward referred to as $CDC2^M$ kinase, in its mitosis-specific form) will include other kinases. The onset of mitosis may therefore be the result of a kinase cascade, at least in part, though exactly what is involved will not become clear until the events of chromosome con-densation, nuclear envelope breakdown and spindle formation are better understood in molecular terms. In the meantime, $CDC2^M$ kinase has been shown to phosphorylate at least two regulatory proteins, namely $pp60^{c-src}$ (Morgan et al. 1989; Shenoy et al. 1989) and c-Abl protein (Kipreos and Wang 1990), both non-receptor tyrosine kinases and proto-oncogenes. Although the physiological consequences of this have yet to be defined, $pp60^{c-src}$ may con-tribute to the rounding up of cells at mitosis since cells transformed by $pp60^{v-src}$ take on a more rounded morphology. It is also noteworthy that some $p34^{cdc2}$ is localised in the centrosome in mammalian cells (Bailly et al. 1989; Riabowol et al. 1989), or its equivalent in yeasts, the spindle pole body (Alfa et al. 1990), a site from which regulation of spindle function is easily envisaged.

Activation of Mitosis-Promoting Factor

Cyclins

The level of $p34^{cdc2}$ remains constant through the cell cycle even though its protein kinase activity rises steeply at the onset of mitosis (Draetta and Beach 1988; Draetta et al. 1988, 1989; Labbé et al. 1988b; Moreno et al. 1989; Simanis and Nurse 1986). It is now well established that this increase depends on the formation of a complex with a class of proteins known as cyclins (Booher et al. 1989; Draetta et al. 1989; Gautier et al. 1990; Labbé et al. 1989a; Meijer et al. 1989; Minshull et al. 1990; Pines and Hunter 1989; Solomon et al. 1990).

The cyclins were first discovered in marine invertebrate eggs as newly syn-thesised proteins whose level fluctuated dramatically following fertilisation (Evans et al. 1983). They accumulated steadily during the cell cycle only to be destroyed abruptly at mitosis. In clams, two cyclins have been detected and isolated, designated A and B, which differ in their kinetics of degradation at mitosis, cyclin A disappearing slightly ahead of cyclin B (Evans et al. 1983; Luca and Ruderman 1989; Swenson et al. 1986; Westendorf et al. 1989). In sea urchin eggs, only a single cyclin has so far been recognised (Pines and Hunt 1987), now known to be B-type (Minshull et al. 1989a). Since their original discovery in marine invertebrates, A- and B-type cyclins have been detected in

Drosophila (Lehner and O'Farrell 1989, 1990; Whitfield et al. 1989), in man (Pines and Hunter 1989, 1990b; Wang et al. 1990), and in *Xenopus* (Minshull et al. 1989b, 1990), the latter containing two B-type cyclins (B1 and B2). A B-type cyclin has also been identified in fission yeast as the product of the *cdc13* gene (Goebl and Byers 1988; Hagan et al. 1988; Solomon et al. 1988). As we will see later, the functions of the two cyclin types appear to overlap, but differences between them are becoming evident.

Since, in many species, most proteins required for early development are pre-made in the oocyte, it seemed reasonable that the comparatively few newly synthesised after fertilisation might perform important functions. The striking periodicity of the cyclins immediately suggested a role in cell cycle regulation, and the first indication that this might be the case came when it was shown that cyclin RNA (both A-type and B-type) injected into *Xenopus* oocytes brought about the activation of MPF and entry into meiotic metaphase (Pines and Hunt 1987; Swenson et al. 1986; Westendorf 1989). Subsequently, elimination of mRNA for cyclin B (both B1 and B2) with antisense oligonucleotides and (endogenous) ribonuclease H (which destroys the resulting duplexes) was found to prevent *Xenopus* egg extracts from entering the mitotic state (Minshull et al. 1989b). This showed that cyclin synthesis is necessary for the normal activation of MPF. In fact, cyclin proved to be the *only* protein required to be synthesised to drive the cell cycle in *Xenopus* eggs (Murray and Kirschner 1989). Thus, (sea urchin) cyclin B RNA alone was sufficient to induce multiple mitotic cycles when added to egg extracts which had been depleted of all endogenous mRNA with pancreatic RNAse (followed by placental RNAse inhibitor to neutralise the added RNAse).

These experiments strongly suggested that cyclins were involved in the activation of the p34^{cdc2} component of MPF, but in themselves provided little direct indication of how this might be accomplished. However, evidence for a physical association between p34^{cdc2} and cyclin is now extensive, making it clear that cyclin is a regulatory subunit of $CDC2^M$ kinase (Pines and Hunter 1990a). For example, p13^{suc1}/Sepharose (which does not interact directly with cyclins) retains both cyclins A and B, in addition to p34^{cdc2}, while monospecific antibodies to cyclin A or B co-precipitate p34^{cdc2}, as well as H1 kinase activity (Booher et al. 1989; Draetta et al. 1989; Giordano et al. 1989; Meijer et al. 1989; Minshull et al. 1990; Pines and Hunter 1989). In addition, the most highly purified preparations of MPF from starfish oocytes contained stoichiometric amounts of both p34^{cdc2} and cyclin B (Labbé et al. 1989a), while highly purified *Xenopus* MPF contained a mixture of cyclins B1 and B2 besides p34^{cdc2} (Gautier et al. 1990).

Phosphorylation and Dephosphorylation of p34^{cdc2}

If the association with cyclin were all that were required to activate p34^{cdc2} then the level of MPF (H1 kinase) would be expected to rise gradually, in parallel with the rise of cyclin. This does not happen; instead, H1 kinase activity increases abruptly at the G$_2$/M transition (Booher et al. 1989; Draetta and Beach 1988; Meijer et al. 1989; Minshull et al. 1990; Moreno et al. 1989; Pines and Hunter 1989). Furthermore, in early embryos, the protein synthesis requirement for division is completed some time before the rise of MPF (Karsenti et al. 1987; Picard et al. 1985; Wagenaar 1983). This indicates that the necessary threshold of cyclin is established well before the increase in MPF actually occurs, which implies the existence of an additional level of regulation. There is now good evidence that this involves the phosphorylation and dephosphorylation of $p34^{cdc2}$.

At the start of the cell cycle, $p34^{cdc2}$ is largely unphosphorylated and monomeric, with the mobility on gel-filtration expected of a 34 kDa protein (Draetta and Beach 1988; D'Urso et al. 1990). As cells progress through S and G_2, $p34^{cdc2}$ becomes more and more phosphorylated, and this is accompanied by a small but significant decrease in its mobility on SDS gel electrophoresis (Draetta and Beach 1988; Draetta et al. 1988; Morla et al. 1989). At the same time, $p34^{cdc2}$ becomes part of a larger complex migrating on gel filtration with an apparent molecular weight of roughly 200 kDa (Brizuela et al. 1989; Draetta and Beach 1988; D'Urso et al. 1990). As already discussed, this complex contains cyclin, but whether there are additional components has not been definitively established since the behaviour of native H1 kinase in different sizing procedures is anomalous, its apparent molecular weight ranging from 200 down to 55 kDa (Arion et al. 1988; Brizuela et al. 1989; Draetta and Beach 1988; Erikson and Maller 1989; Labbé et al. 1989a; Pondaven et al. 1990). Nevertheless, the 200 kDa complex most likely represents a pair of cyclin/$p34^{cdc2}$ dimers. Significantly, cyclin B is associated only with the most phosphorylated derivatives of $p34^{cdc2}$ (Brizuela et al. 1989; Draetta and Beach 1988; D'Urso et al. 1990), though the situation with cyclin A is less clear cut (Giordano et al. 1989; Pines and Hunter 1990).

Phosphorylation of $p34^{cdc2}$ is on serine, to some degree, but mainly on threonine and tyrosine (Draetta et al. 1988; Gould and Nurse 1989; Morla et al. 1989). Indeed, $p34^{cdc2}$ is the most abundant phosphotyrosine-containing protein in HeLa cells (Draetta et al. 1988). The pattern of phosphorylation is actually complex since on two-dimensional gels, $p34^{cdc2}$ gives up to 12 distinct spots, at least 7 of which are eliminated by phosphatase treatment (Giordano et al. 1989). (The basis of the difference between the remaining five spots is unclear). Nevertheless, despite the complexity, it is clear that activation to generate a functional H1 kinase involves substantial de-phosphorylation on both tyrosine and threonine (Dunphy and Newport 1989; Gautier et al. 1989; Gould and Nurse 1989; Labbé et al. 1989b; Morla et al. 1989; Pondaven et al. 1990). Treatment with vanadate (a tyrosine phosphatase inhibitor) prevents kinase activation, underscoring the particular importance of tyrosine dephosphorylation (Morla et al. 1989; Solomon et al. 1990). In fission yeast, the tyrosine residue phosphorylated has been identified as Tyr_{15} (Gould and Nurse 1989). This is located in the ATP binding site, and phosphorylation here would be expected to prevent ATP binding, so blocking kinase activity. Indeed, mutation of Tyr_{15} to phenylalanine brings about premature activation of the kinase and advancement of mitosis such that cells divide at a fraction of their normal size (the "wee" phenotype) (Gould and Nurse 1989). In mammalian cells, the adjacent Thr_{14} is also phosphorylated giving an additional level of control (cited in Solomon et al. 1990), and treatment with a purified tyrosine phosphatase is insufficient for kinase activation (Morla et al. 1989).

Not all of the phosphate is removed from $p34^{cdc2}$ at the G_2/M transition (Solomon et al. 1990) and there is increasing evidence that what remains (possibly involving Thr_{167} – Hunt 1991; Lewin 1990) is necessary for H1 kinase activation (see below). The fact that it depends on both phosphorylation as well as dephosphorylation helps to explain why H1 kinase activation, though reported (Pondaven et al. 1990), is difficult to achieve in vitro by phosphatase treatments (Lewin 1990).

In addition to activating phosphorylations of $p34^{cdc2}$, the cyclin subunit also becomes phosphorylated, coincident with H1 kinase activation (Gautier et al. 1990; Meijer et al. 1989; Minshull et al. 1989a; Patel et al. 1989; Pondaven et al. 1990). Since cyclin is a good substrate for the activated $p34^{cdc2}$ kinase in vitro (Draetta and Beach 1988; Gautier et al. 1990; Minshull et al. 1990; Pines and Hunter 1989), its phosphorylation in vivo is most likely brought about by the $p34^{cdc2}$ to which it is complexed, though this is not yet proven. The consequences of cyclin phosphorylation are far from clear. Given that a small amount of active MPF is able to activate a latent pool of pre-MPF in *Xenopus* oocytes, the autophosphorylation of the cyclin subunit is an attractive explanation of autoactivation (Brooks 1989). However, there is little direct evidence to support this. Indeed, Kirschner and colleagues have identified an inhibitor of MPF activation in *Xenopus* oocytes which, on purification, proved to be a type-2A phosphatase (Lee et al. 1991). This removes phosphate from both cyclin and $p34^{cdc2}$, and directly inhibits the activated H1 kinase. However, the kinetics of inactivation correlate more closely with loss of phosphate from $p34^{cdc2}$ than from cyclin (Lee et al. 1991).

Because cyclin is associated with the most highly phosphorylated forms of $p34^{cdc2}$, it was originally presumed that phosphorylation preceded binding. In fact, recent evidence suggests the converse. In cell-free extracts of *Xenopus* eggs in which cyclin synthesis was blocked with cycloheximide, no phosphorylation of $p34^{cdc2}$ occurred *until* the addition of exogenous cyclin (Solomon et al. 1990). This strongly suggests that it is the association with cyclin which drives $p34^{cdc2}$ phosphorylation. The inhibitory phosphorylations of $p34^{cdc2}$ are therefore deemed to be part of the activation pathway, designed to prevent premature activation of the kinase until an appropriate level of the inactive cyclin/$p34^{cdc2}$ complex (pre-MPF) has accumulated. Subsequent removal of the inhibitory phosphorylations then permits the sudden and massive rise of kinase activity responsible for triggering mitosis.

Although the precise order in which the various phosphorylations and dephosphorylations take place has yet to be defined, a scheme for the activation of $p34^{cdc2}$ consistent with current information is shown in Fig. 1.1a. On association with cyclin, $p34^{cdc2}$ becomes the target for so far unidentified tyrosine and threonine kinases which introduce inhibitory phosphorylations on Tyr_{15} and Thr_{14}, and possibly elsewhere. Another threonine kinase introduces activating phosphorylations (possibly involving Thr_{167}) to generate the inactive, but "primed" complex known as pre-MPF. Though not proven, it seems likely that this kinase requires the inactivated form of $p34^{cdc2}$ (phosphorylated on Tyr_{15} and Thr_{14}) as substrate (Hunt 1991). These activating and inhibiting phosphorylations are opposed by a corresponding battery of phosphatases. The type-2A phosphatase known as INH, identified by Lee et al. (1991), removes the activating phosphorylation(s) while unidentified tyrosine and threonine phosphatases remove the inhibitory phosphorylations on Tyr_{15} and Thr_{14}, a step crucial to the eventual activation of MPF/H1 kinase.

The autoactivation of MPF may be understood at least in part following the demonstration that activation is associated with a substantial increase in the

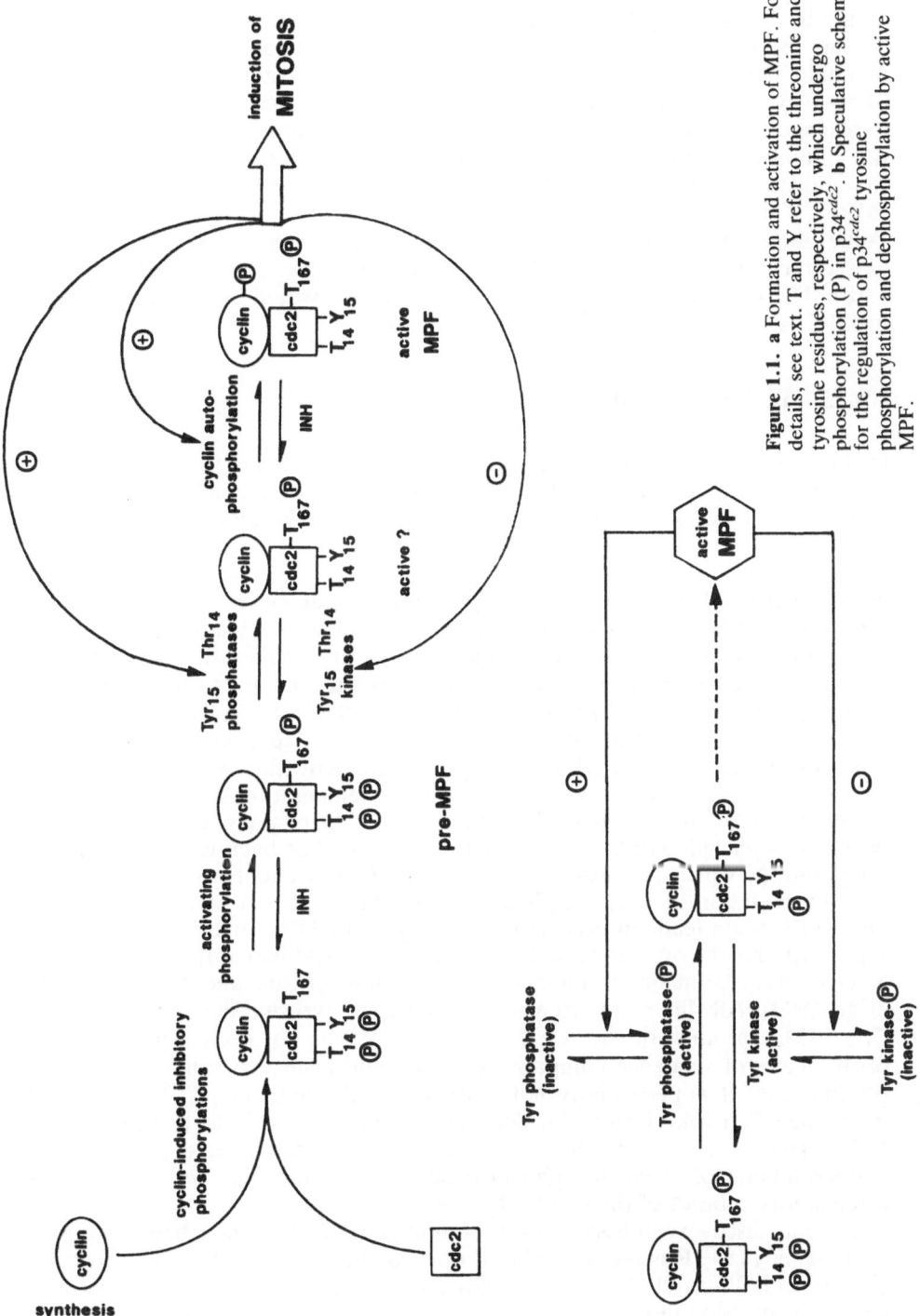

Figure 1.1. a Formation and activation of MPF. For details, see text. T and Y refer to the threonine and tyrosine residues, respectively, which undergo phosphorylation (P) in p34^{cdc2}. b Speculative scheme for the regulation of p34^{cdc2} tyrosine phosphorylation and dephosphorylation by active MPF.

rate of tyrosine and threonine dephosphorylation, together with a large decrease in the rate of tyrosine re-phosphorylation (Solomon et al. 1990). It may be supposed that as cyclin accumulates, the equilibria are initially such as to maintain the complex with $p34^{cdc2}$ in its fully inactivated (Tyr_{15} and Thr_{14} phosphorylated) state. As the amount of pre-MPF increases however, a point is reached when the ability of the tyrosine and threonine kinases to maintain full inhibition in the face of phosphatase activity is exceeded. The result is the appearance of a small amount of dephosphorylated and active H1 kinase (MPF). This in turn might be supposed to activate fully the tyrosine and threonine phosphatases and simultaneously inhibit the opposing kinases, perhaps as a direct (or indirect) result of phosphorylating the relevant enzymes (Fig. 1.1b). The consequences of such positive feedback would be an explosive rise in MPF/H1 kinase activity.

cdc25 and the Timing of MPF Activation

In the scheme outlined in Fig. 1.1a, the accumulation of cyclin is seen to drive the activation of MPF. Although this is compatible with the cell cycles of early embryos, it is not a sufficient explanation of the regulation of the G_2/M transition in most other situations. In fission yeast, for example, no evidence has been obtained implicating expression of cdc13 (a B-type cyclin) in the timing of cell division (Moreno et al. 1989). Similarly, the onset of asynchrony in Drosophila embryos after mitosis 14 is not accompanied by any corresponding variation in the accumulation of cyclins (Lehner and O'Farrell 1989, 1990).

In fission yeast, there is good evidence that the timing of mitosis depends on the product of the cdc25 gene, a dosage-dependent, positive regulator of $p34^{cdc2}$ (Nurse 1990). In the absence of cdc25 function, cells fail to enter mitosis, but over-expression of cdc25 leads to premature activation of $p34^{cdc2}$ and to a wee phenotype. In cells arrested in G_2 by a temperature-sensitive mutation in cdc25, $p34^{cdc2}$ is hyperphosphorylated on both tyrosine and threonine, but becomes dephosphorylated within minutes of the shift back to the permissive temperature, which restores cdc25 function (Gould and Nurse 1989). Significantly, mutation of Tyr_{15} to phenylalanine, which cannot be phosphorylated and which again leads to precocious activation of $p34^{cdc2}$ and to a wee phenotype, bypasses the need for cdc25 function (Gould and Nurse 1989). Moreover, a defect in cdc25 can be complemented by a human tyrosine phosphatase (Gould et al. 1990). All this points to a role for cdc25 in tyrosine dephosphorylation. The 80 kDa product of the cdc25 gene ($p80^{cdc25}$) is not itself a phosphatase, on the basis of sequence comparisons, but could instead be an activator of a phosphatase. It is particularly noteworthy that the level of $p80^{cdc25}$ increases more than four-fold during G_2, decreasing again after mitosis (Moreno et al. 1990). The cdc25 gene product may therefore be a cell cycle dependent regulatory subunit of a tyrosine phosphatase, in much the same way as cyclin is a regulatory subunit of the $p34^{cdc2}$ kinase.

Since its discovery in fission yeast, homologues of cdc25 have been identified in budding yeast (Russell et al. 1989), man (Sadhu et al. 1990) and Drosophila (Edgar and O'Farrell 1989). In the latter case, regional variations in the level of gene expression in the embryo after mitosis 14 correlate well with the timing of division in the subsequently asynchronous cell cycles. Involvement of cdc25 homologues in establishing the timing of $p34^{cdc2}$ activation may therefore be universal.

Dependence of Mitosis on DNA Synthesis

In most cell types, inhibition of DNA synthesis prevents the initiation of mitosis. It has recently been established in fission yeast that this important control is mediated through the *cdc25* gene product (Enoch and Nurse 1990). Cells over-expressing $p80^{cdc25}$ fail to arrest mitosis when DNA synthesis is blocked, as do cells bearing mutant alleles of *cdc2* that bypass the need for *cdc25*. Given that its level normally rises during G_2, the most likely basis for the control is that the accumulation of $p80^{cdc25}$ is not permitted until DNA replication has been completed, though changes in activity (as distinct from abundance) are not ruled out. In budding yeast, the *RAD9* gene product is required to mediate the inhibition of mitosis when DNA replication is blocked (Weinert and Hartwell 1989) and presumably acts through the budding yeast homologue of *cdc25* (*MIH1*). Other clues come from mammalian cells in which exposure to high levels of caffeine uncouples the link between DNA synthesis and mitosis such that cells continue to divide in the presence of DNA synthesis inhibitors (Schlegal and Pardee 1986). However, the precise target of caffeine is unknown.

Although mitosis is dependent on the completion of DNA synthesis in most cell types, this is not the case during the embryonic cleavage cycles of many invertebrates and amphibians. In *Drosophila* embryos there is a large pool of maternal RNA for *string*, the homologue of *cdc25*, which is abruptly destroyed at the start of cycle 14 (Edgar and O'Farrell 1989). Thereafter, mitosis is dependent on zygotic transcription of *string* and, as already mentioned, the pattern of expression is a predictor of the timing of mitosis. The high level of *string* RNA, like over-expression of *cdc25* in fission yeast, explains why mitotic cycles similarly continue in the presence of DNA synthesis inhibitors (Raff and Glover 1988). Though not yet proven, it is likely that *Xenopus* eggs have large stores of maternal cdc25 protein, again accounting for the lack of dependence of mitosis on DNA synthesis (Kimelman et al. 1987). The timing of mitosis would then be dictated solely by the accumulation of cyclin rather than by the need to synthesise new cdc25 protein.

The *wee1* and *nim1* Genes

In fission yeast, two other genes involved in controlling the timing of mitosis are *wee1*, a negative regulator of $p34^{cdc2}$ and *nim1*, an inhibitor of *wee1* (Nurse 1990). Sequence comparison suggests that *nim1* codes for a serine/threonine protein kinase, the target of which could well be the *wee1* gene product itself. Certainly, it acts to switch off *wee1* function. The *wee1* gene also codes for a protein kinase with a novel specificity for serine and tyrosine (Featherstone and Russell 1991). This immediately raises the question of whether *wee1* is the tyrosine kinase responsible for the inhibitory phosphorylation of $p34^{cdc2}$ on tyrosine. Unfortunately, this seems unlikely, since tyrosine phosphorylation of $p34^{cdc2}$ still occurs when *wee1* function is eliminated (Gould et al. 1990). Another potential target might be the activation of INH (Fig. 1.1a), but many other possibilities exist. Elucidation of the mechanism by which the *wee1* product inhibits $p34^{cdc2}$ is clearly a matter of some importance.

The *suc1* Gene

As already mentioned, the *suc1* gene was identified as a suppressor of certain *cdc2* mutations (Hayles et al. 1986a,b). The two gene products interact physically (Brizuela et al. 1987) and $p13^{suc1}$/Sepharose is in widespread use as a specific affinity matrix for the isolation of $p34^{cdc2}$. It is therefore ironic that the function of $p13^{suc1}$ in vivo remains unclear. It has been suggested that $p13^{suc1}$ is a subunit of the $p34^{cdc2}$ kinase, along with cyclin, on the basis of co-immuno-precipitation of $p13^{suc1}$ with antibodies to $p34^{cdc2}$ (Draetta and Beach 1988; Draetta et al. 1988; Giordano et al. 1989). However, purified *Xenopus* MPF has not been reported to contain components of a size corresponding to $p13^{suc1}$ (Lohka et al. 1988). In addition, it is difficult to detect $p34^{cdc2}$ in immuno-precipitates of $p13^{suc1}$ (Brizuela et al. 1987; Draetta and Beach 1988). This suggests that only a minor fraction of $p13^{suc1}$ is complexed to $p34^{cdc2}$ in vivo, despite the fact that fission yeast cells, at least, contain five times more $p34^{cdc2}$ than $p13^{suc1}$ (Brizuela et al. 1987). In fission yeast, deletion of *suc1* leads to arrest in mitosis, with higher levels of H1 kinase activity than normal (Moreno et al. 1989), whereas addition of $p13^{suc1}$ to *Xenopus* egg extracts prevents the activation of $p34^{cdc2}$ and the associated tyrosine dephosphorylation (Dunphy and Newport 1989). Since $p13^{suc1}$ does not inhibit $p34^{cdc2}$ kinase directly, once activated, these observations suggest that $p13^{suc1}$ may be required to prevent precocious activation of $p34^{cdc2}$ and to assist in its inactivation after mitosis. However, the way in which this is accomplished is unknown.

Calcium

So far, the timing of mitosis has been discussed in terms of the accumulation of cyclin and $p80^{cdc25}$ and in terms of the cycles of $p34^{cdc2}$ phosphorylation/dephosphorylation they help regulate. However, for many cell types, the ultimate trigger for $p34^{cdc2}$ activation (i.e. tyrosine and threonine dephosphorylation) may actually be a transient rise in intracellular free Ca^{2+} (Whitaker and Patel 1990). In sea urchin embryos, a Ca^{2+} transient immediately precedes nuclear envelope breakdown, and blocking this transient with Ca^{2+} chelators such as EGTA or BAPTA prevents the onset of mitosis (Steinhardt and Alderton 1988; Twigg et al. 1988). Conversely, artificially raising the Ca^{2+} concentration can induce premature entry into mitosis, provided a certain minimum amount of new protein synthesis (presumably cyclin) has occurred (Steinhardt and Alderton 1988; Twigg et al. 1988). Similar conclusions have been reached for mouse 3T3 cells (Kao et al. 1990). Despite all this, cell-free extracts of *Xenopus* eggs undergo repeated mitotic cycles in the presence of high levels of EGTA (cited in Whitaker and Patel 1990). This has been taken to imply that Ca^{2+} transients are not essential for the activation of $p34^{cdc2}$ kinase. Nevertheless, the evidence in favour of a role for Ca^{2+} in intact cells is compelling. It is possible that the *Xenopus* cell-free system does not fully reproduce in vivo behaviour, the accumulation of cyclin eventually driving $p34^{cdc2}$ activation even in the absence of a normal Ca^{2+} transient. It is also possible that *Xenopus* eggs differ from other cells in their lack of dependence on a Ca^{2+} signal for the initiation of mitosis. For instance, in cleaving *Xenopus* embryos, Rink et al. (1980) were unable to detect Ca^{2+} transients during the cell cycle using Ca^{2+}-selective microelectrodes, while buffering intracellular free Ca^{2+} to 0.1 μM with EGTA failed to prevent cell division. It is to be hoped that these differences between *Xenopus* and other species will soon be resolved. As for how Ca^{2+} might provoke the activation of $p34^{cdc2}$ in those cell types in which its role is established, there is now good evidence for the involvement of the multifunctional Ca^{2+}/calmodulin-dependent protein kinase (CaM kinase) (Baitinger et al. 1990). How this impinges on the regulatory circuit outlined in Fig. 1.1a is not yet clear.

Why Two Classes of Cyclin?

Until now, the functions of the cyclin A and cyclin B complexes with $p34^{cdc2}$ have been tacitly taken to be similar, but it is becoming increasingly apparent that this is not strictly the case.

The two classes of cyclins (A-type and B-type) share a certain degree of homology over a central region of roughly 200 amino acids (the so-called cyclin box) but are otherwise highly divergent. Even within the cyclin box, the similarity between classes is less than between members of the same class, and there are particular sequence motifs which serve to distinguish between the two (Minshull et al. 1989a, 1990).

Both cyclin types associate with $p34^{cdc2}$, and the complexes have H1 kinase activity (Draetta et al. 1989; Giordano et al. 1989; Minshull et al. 1990; Pines and Hunter 1990b). In addition, both cyclins are able to induce *Xenopus* oocyte maturation (Pines and Hunt 1987; Swenson et al. 1986; Westendorf et al. 1989). Conversely, synthesis of cyclin B alone (Murray and Kirschner 1989) or cyclin A alone (T. Hunt, personal communication) is sufficient for the induction of mitosis in *Xenopus* egg extracts depleted of endogenous mRNA. All this suggests an equivalence of function. Nevertheless, sequence comparisons indicate that both classes have been conserved independently during evolution. Furthermore, the two types are regulated independently, cyclin A accumulating earlier in the cycle and disappearing earlier in mitosis than cyclin B (Evans et al. 1983; Luca and Ruderman 1989; Minshull et al. 1990; Pines and Hunter 1990b). The disappearance of cyclin B, which normally coincides with the metaphase/anaphase transition, is prevented by colchicine or nocodazole, whereas the disappearance of cyclin A is unaffected (Minshull et al. 1989a; Pines and Hunter 1990). Such behaviour argues for distinctive functions and, in support of this, elimination of the gene for cyclin A causes cell cycle arrest in *Drosophila* embryos after mitosis 15, once the pool of maternal cyclin A protein is exhausted (Lehner and O'Farrell 1989, 1990). Evidently, cyclin A must fulfil some role that cannot easily be duplicated by cyclin B (which continues to be expressed normally), in marked contrast to the results with *Xenopus* egg extracts (Murray and Kirschner 1989). It is possible that the cell-free system does not fully reproduce in vivo behaviour, perhaps because the exogenous cyclins are accumulated to higher levels than normal. Consistent with such quantitative considerations, it has been reported that injection of a mixture of A- and B-type cyclin RNA into *Xenopus* oocytes is more effective at inducing maturation than an equivalent amount of either type alone (Westendorf et al. 1989), though such synergy is apparently not observed reproducibly (T. Hunt, personal communication).

Monomeric p34^{cdc2} has weak kinase activity towards casein as substrate and only becomes able to phosphorylate histone H1 after association with cyclin (Brizuela et al. 1989; Draetta and Beach 1988). This suggests that the cyclin subunit may function to direct the p34^{cdc2} catalytic subunit towards appropriate substrates. It is therefore interesting that p34^{cdc2}/cyclin A and p34^{cdc2}/cyclin B complexes, though both active towards histone H1, show significant differences in their dependence on histone H1 concentration, as well as marked differences in their preference for other substrates (Minshull et al. 1990). The cyclin A and cyclin B complexes are therefore not equivalent but have different, albeit overlapping, substrate specificities.

A further function of the cyclin subunit may be to direct the p34^{cdc2} catalytic subunit to particular locations within the cell. In fission yeast, cyclin B apparently targets p34^{cdc2} to the nucleus (Booher et al. 1989). By contrast, in human cells, cyclin B is confined in the cytoplasm until prophase, whereas cyclin A accumulates in both cytoplasm and nucleus throughout interphase (Pines and Hunter 1990a). It is noteworthy that in human cells, the H1 kinase activity of the p34^{cdc2}/cyclin A complex also rises earlier and much less abruptly than that of the cyclin B complex, beginning during S phase rather than at the onset of mitosis (Pines and Hunter 1990b). This, coupled with the partly nuclear location, raises the possibility that the p34^{cdc2}/cyclin A complex may have some role in DNA replication in addition to mitosis (see later). Another possibility is that the cyclin A complex is involved in chromosome condensation. To some degree, condensation begins in S phase as each chromosomal segment is replicated, and continues progressively through G$_2$ to prophase, at which point the chromosomes finally become visible (Mullinger and Johnson 1983).

Finally, although cyclin A undoubtedly complexes with p34^{cdc2} to generate H1 kinase activity (Minshull et al. 1990), in human cells it is apparently associated mainly with a related but different protein (Pines and Hunter 1990b). This binds to p13^{suc1}, like p34^{cdc2}, and is recognised by some (but not all) antibodies to p34^{cdc2}, but its apparent molecular weight is slightly less (32 kDa), and its tryptic peptide map is distinct. The complex formed between this *cdc2*-related protein and cyclin A is also reported to have H1 kinase activity. In *Xenopus*, a possibly similar protein has been identified, known as Eg1 (Paris et al. 1991). This does not associate with cyclin B (Solomon et al. 1990), but it is not known whether it interacts with cyclin A. At present, the significance of these *cdc2*-related proteins is unclear, but their existence raises the possibility that the M-phase H1 kinase activity currently attributed exclusively to p34^{cdc2} may actually be the result of a small family of related kinases. In yeasts, the supremacy of *cdc2* (or its homologue in budding yeast, *CDC28*) is clear from genetics, but for higher eukaryotes it is possible that there has been some duplication of *cdc2* function.

Exit from Mitosis

Cyclin Degradation

The consequences of p34^{cdc2} kinase activation include not only entry into mitosis (i.e. chromosome condensation, nuclear envelope breakdown, spindle formation, etc.) but also induction of cyclin proteolysis. The degradation of cyclin which ensues leads to the inactivation of p34^{cdc2} and is essential for exit from the mitotic state. This has been established beyond doubt following the discovery that a truncated cyclin lacking the first 90 amino acids was resistant to degradation (Murray et al. 1989). It was nevertheless fully able to activate the p34^{cdc2} kinase, but in the absence of degradation, the resulting mitotic state was maintained indefinitely.

Comparison of different cyclins reveals little sequence conservation in the first 100–150 amino acids, with the exception of a small island of 9 amino acids corresponding to residues 42–50 in sea urchin cyclin B. This segment, now referred to as the "destruction box", and an adjacent region rich in lysines, together are necessary and sufficient for mitosis-specific degradation (Glotzer et al. 1991). Moreover, fusion of an N-terminal domain of cyclin containing these elements to heterologous proteins renders them similarly unstable at mitosis. By adding such constructs to *Xenopus* egg extracts arrested in mitosis by an N-terminal truncated cyclin, Glotzer et al. (1991) were able to show that degradation was preceded by poly-ubiquitination, most probably involving some or all of the lysines located C-terminal to the destruction box. Ubiquitin is a small, highly conserved protein (molecular weight 7000) involved in marking proteins for destruction, and the resulting ubiquitinated cyclins or cyclin fusion proteins become targets for proteolysis. The ligase responsible for ubiquitin conjugation has not yet been identified but presumably recognises elements of the destruction box. It is also presumably activated, most probably indirectly, by active $CDC2^M$ kinase (Felix et al. 1990), though the details remain to be worked out.

Regulation of Cyclin Destruction by the Mitotic Spindle

Although active $CDC2^M$ kinase induces cyclin degradation, thereby terminating its own activity and permitting exit from mitosis, in most cell types this negative feedback loop is regulated by some aspect of spindle function. With the exception of the early embryonic cycles of some species (e.g. *Xenopus*, *Drosophila*), the completion of mitosis is not permitted until all chromosomes are fully engaged with the mitotic spindle, whereupon some signal (most probably the destruction of cyclin B) triggers the metaphase/anaphase transition. The existence of such a control is clearly seen when the spindle is disrupted by microtubule inhibitors such as nocodazole or colchicine. Such treatments prolong the mitotic state and prevent the destruction of cyclin B (Minshull et al. 1989; Pines and Hunter 1990). As already noted, however, degradation of cyclin A proceeds on schedule. The structures best placed to detect unengaged chromosomes are, of course, the kinetochores, but as yet there is no direct evidence implicating them in regulating the destruction of cyclin B.

Cytostatic Factor and the c-*mos* Proto-oncogene

Xenopus eggs are naturally arrested in metaphase of the second meiotic division as a result of stabilisation of $CDC2^M$ kinase (and its associated cyclin B). The activity responsible for this, termed cytostatic factor or CSF (Masui and Markert 1971), has recently been shown to be the product of the c-*mos* proto-oncogene (Sagata et al. 1989a; Hunt 1989b), a serine/threonine protein kinase. Cyclin B is a substrate of the c-*mos* kinase in vitro (Roy et al. 1990), but it is not yet known whether this is relevant to its stabilisation. The other obvious possibility is that the c-*mos* kinase inhibits the cyclin-specific ubiquitin ligase.

Synthesis of c-*mos* protein in *Xenopus* occurs only during progesterone-induced oocyte maturation, and is essential for normal maturation (presumably by stabilising the active configuration of $CDC2^M$ kinase). Thus, injection of antisense oligonucleotides to c-*mos* blocks progesterone-induced maturation (Sagata et al. 1988), whereas injection of c-*mos* RNA, like that for cyclin, induces maturation in the absence of progesterone (Sagata et al. 1989b). On fertilisation, c-*mos* protein is destroyed, probably by the calcium-dependent protease calpain (activated by the Ca^{2+} transient following fertilisation), and never again accumulates to the levels found in the egg (Watanabe et al. 1989). This might suggest that c-*mos* functions solely in meiosis. However, low levels of expression during mitotic cycles are not excluded, and it is possible that c-*mos* protein plays some part in stabilising cyclin until all chromosomes are attached to the spindle. At the very least, the mechanism by which c-*mos* stabilises cyclin may shed some light on how the latter control operates.

Given its capacity to arrest mitosis, it is difficult to understand why the c-*mos* protein should have transforming potential. Moderate levels of expression are indeed growth inhibitory (Papkoff et al. 1982), as one might now expect, though there is not yet evidence for arrest in mitosis. At present, it is unclear why low levels of expression should lead to uncontrolled proliferation. One attractive possibility (Hunt 1989b) is that c-*mos* stabilises hypothetical G_1 cyclins believed to be involved in regulating the G_1/S transition (see later).

Cell Cycle Initiation

Role of *cdc2*

As indicated at the outset, the *cdc2* gene of fission yeast, and its counterpart, *CDC28*, in budding yeast, are required not only at mitosis but also for the initiation of the cell cycle at START. Given the high degree of conservation of mitotic mechanisms, it would be surprising if p34^{cdc2} were not also involved in regulating cell cycle initiation in higher eukaryotes, and direct evidence for this has come recently from several directions.

Cell-free extracts of *Xenopus* eggs are able to support cycles of DNA replication as well as mitosis (Blow et al. 1989; Hutchison et al. 1989). Chromatin, or even naked DNA, added to such extracts is first assembled into nucleus-like structures, complete with double envelope and nuclear pores. DNA synthesis then follows in a cell cycle-dependent fashion, in that replication is restricted to a single round until the extracts pass through the mitotic state (under the influence of $CDC2^M$ kinase), after which a further round of replication becomes possible. Extracts depleted of p34^{cdc2} with either p13^{suc1}/Sepharose or antibodies to p34^{cdc2}, remain fully capable of nuclear assembly, as well as the continuation of DNA synthesis once started. They are, however, prevented from *initiating* DNA synthesis (Blow and Nurse 1990). This firmly establishes a role for p34^{cdc2}, or a very closely related molecule, in a step required for the initiation of DNA synthesis. Interestingly, this step is completed some time before DNA replication actually begins, since depletion of p34^{cdc2} later than 15 min after adding chromatin to the extracts fails to block initiation even though replication does not begin until 45 min. Whether this step is equivalent to START (i.e. the rate-limiting event for cell cycle initiation) in the intact cell remains to be established. In the meantime, the ability to restore the capacity to initiate DNA synthesis by adding back material eluted from the p13^{suc1}/Sepharose using soluble p13^{suc1}(Blow and Nurse 1990), should enable definition of the form of p34^{cdc2} involved. It is obviously of critical importance to determine how this differs from the form required at mitosis.

Evidence for the involvement of p34^{cdc2} in the initiation of DNA synthesis has also come from studies of SV40 DNA replication in vitro. SV40 is a small DNA virus which relies almost entirely on host proteins for replication, apart from a requirement for the virally encoded T antigen at initiation (Fairman et al. 1989). Cell-free extracts prepared from S-phase cells, supplemented with T antigen, are able to support viral DNA replication, but extracts from G_1 cells are not (Roberts and D'Urso 1989). A factor has been partially purified from S-phase cells (RF-S) which overcomes the deficiencies of G_1 extracts, enabling them now to support viral DNA replication. This factor co-purifies with H1 kinase activity and contains p34^{cdc2} (D'Urso et al. 1990). The association is not coincidental since passage over p13^{suc1}/Sepharose eliminates both p34^{cdc2} and RF-S activity.

A need for p34^{cdc2} at the G_1/S transition has also been demonstrated for mitogen-stimulated T lymphocytes (Furukawa et al. 1990). The level of p34^{cdc2} does not change during the cell cycle of growing cells but falls when cells become quiescent (Morla et al. 1989). Freshly isolated T cells have very little p34^{cdc2} (presumably because of prolonged quiescence), but the level rises dramatically just before S phase. Blocking this increase with antisense oligonucleotides to *cdc2* prevents the onset of DNA synthesis, an affect not expected if p34^{cdc2} acts only at mitosis (Furukawa et al. 1990).

G_1 Cyclins

As we have seen, entry into mitosis depends upon the association between p34^{cdc2} and other polypeptides (notably cyclins). It is probable that similar mechanisms are involved in regulating the G_1/S transition. Certainly, p34$^{cdc2/28}$ is monomeric in G_1 but becomes part of a high molecular weight complex in S phase (Draetta and Beach 1988; D'Urso et al. 1990; Wittenberg and Reed 1988), though exactly when complex formation occurs relative to the G_1/S transition is less certain, given the limitations of cell synchronisation procedures.

In budding yeast, three important regulators of the G_1/S function of *CDC28* have been uncovered. Two, *CLN1* and *CLN2*, are closely related and were isolated as dosage-dependent suppressors of certain temperature sensitive *CDC28* alleles (Reed et al. 1989). The third, known variously as *WHI1* or *DAF1* (Cross 1989; Nash et al. 1988), but recently renamed *CLN3* (Richardson et al. 1989), is less closely related, and was identified through dominant mutations which accelerate progress through G_1. As a result of the mutation, cells reach division at half the size of wild-type cells. (This is similar to the wee phenotype of fission yeast, but the mutations here act at G_1/S rather than at G_2/M.) Dominant mutations in *CLN1* and *CLN2* have also been identified which similarly accelerate entry into S phase, resulting in reduced size at division (Hadwiger et al. 1989). The three gene products are thus strongly implicated in regulating the timing of the G_1/S transition, and evidence for physical interaction with the *CDC28* protein has recently been obtained (Wittenberg et al. 1990). Significantly, all three genes show homology to the cyclin family, but less than between previously identified members of the group (A-type or B-type). They may therefore define a new sub-family of cyclin-like proteins concerned more with the G_1/S transition than with entry into mitosis. In keeping with this, *CLN1* and *CLN2* proteins have been shown to accumulate in G_1 and to disappear after entry into S phase (Wittenberg et al. 1990). They also disappear in response to mating pheromones which bring about arrest in G_1.

A feature common to *CLN1*, *CLN2* and *CLN3* is the presence of so-called PEST sequences, characteristic of unstable proteins, near their C termini. Significantly, these sequences are deleted in the dominant mutations which advance the G_1/S transition (Cross 1988; Hadwiger et al. 1989; Nash et al. 1988). Such deletions would be expected to stabilise the proteins, which in turn suggests that the accelerated passage through G_1 may be due to premature accumulation to a critical threshold. These mutations are reminiscent of the N-terminal deletions that remove the destruction box from A- and B-type cyclins, preventing degradation at mitosis. Though not proven, an attractive possibility is that changes in the stability of the *CLN* proteins are involved in monitoring environmental conditions. It is easy to imagine that they might be sufficiently stable under optimal conditions to accumulate to levels necessary for the activation of *CDC28* kinase, but become rapidly unstable when conditions deteriorate, leading to arrest in G_1. It is possible that the three *CLN* proteins are involved in monitoring different aspects of the environment, but their functions evidently overlap since all three genes must be eliminated to completely block the cell cycle, i.e. a single *CLN* gene is sufficient to permit entry into S phase (Richardson et al. 1989).

So far, homologues of the *CLN* genes have not been identified in organisms other than budding yeast. Conversely, A-type cyclins have not yet been found in budding yeast. Attention has already been drawn to the lack of clear insight into the precise role of cyclin A. Although cyclin A can undoubtedly trigger entry into M phase when ectopically expressed (Swenson et al. 1986; Westendorf et al. 1989), it is puzzling that the H1 kinase activity of p34^{cdc2}/cyclin A complexes should normally rise during S phase and fall significantly before the metaphase/anaphase transition (Giordano et al. 1989; Minshull et al. 1989a, 1990; Pines and Hunter 1990). Given such behaviour, it is worth considering the possibility that cyclin A in higher eukaryotes may actually be involved in regulating the G_1/S transition and be a counterpart of the *CLN* proteins of budding yeast. This is not entirely groundless. In the first place, addition of purified clam cyclin A to extracts of (human) G_1 cells is sufficient to activate SV40 DNA replication to levels seen in S-phase extracts (D'Urso et al. 1990). Cyclin A therefore induces the appearances of replication factor RF-S which, as already discussed, is associated with $p34^{cdc2}$. Secondly, in cells infected with or transformed by adenovirus, cyclin A is found complexed to the viral transforming protein, E1A (Giordano et al. 1989; Pines and Hunter 1990). This association may be mere coincidence and of no consequence, but a simi!ɔr association between E1A and the retinoblastoma gene product, Rb, is believed to be important for cell transformation (Weinberg 1989). Another connection with transformation comes from the discovery of hepatitis B virus integrated next to the cyclin A gene in a primary liver tumour (Wang et al. 1990). As before, this may be no more than coincidence – the virus has to integrate somewhere – but the parallels with proto-oncogene activation by retroviral insertion (Peters 1989) are striking. With both E1A and hepatitis B virus, the association with transformation, if not coincidence, is difficult to understand if cyclin A acts only at mitosis, given that mammalian cell proliferation is regulated mainly in G_1. A role at the G_1/S transition would make much more sense.

If cyclin A can activate the G_1/S function of $p34^{cdc2}$ in vertebrate cells, it is clearly not the only such factor. Immediately after activation, *Xenopus* eggs contain no detectable cyclin A (T. Hunt, personal communication), yet are capable of initiating and completing DNA replication in the presence of cycloheximide, which would prevent new cyclin A synthesis (Newport and Kirschner 1984). The same is true of cell-free extracts prepared from such eggs (Blow and Laskey 1988). It follows that there must be some other activator of $p34^{cdc2}$ present constitutively in egg cytoplasm, independent of new protein synthesis. The identity of this factor is clearly a matter of some importance.

Conclusion

Irrespective of the precise role of cyclin A, given the discovery of G_1 cyclins in budding yeast, it will surely be only a matter of time before comparable proteins are identified in higher eukaryotes. The physical association between the *CLN* proteins and $p34^{CDC28}$ (Wittenberg et al. 1990) strongly suggests that passage of START depends on kinase activation, in much the same way as does the initiation of mitosis. Nevertheless, entry into S phase is not accompanied by anything like the burst of protein phosphorylation which heralds the onset of mitosis. What might be called the $CDC2^S$ kinase must therefore be much more restricted in its choice of substrates, or more highly localized within the cell, than its M-phase counterpart.

A further complication in attempting to define the role of the $CDC2^S$ kinase is that the nature of START is still by no means obvious. In budding yeast, a number of genes have been identified as acting between *CDC28* and the initiation of DNA synthesis (Pringle and Hartwell 1982), but until their functions are clarified, it remains possible that a cell is not committed to divide until the first nucleotides are incorporated into DNA. Be that as it may, the initiation of DNA synthesis continues to be the earliest unambiguous indication of the passage of START.

Another aspect of START that must be accommodated is the extreme variability in G_1, even within a population of otherwise identical cells (Brooks 1985). It is conceivable that this variability reflects differences in the time taken to accumulate G_1 cyclins to the necessary threshold, though why the rates should vary is unclear. Alternatively, activation of the $CDC2^S$ kinase may be inherently variable even after G_1 cyclins have attained an adequate level. The elements of positive feedback involved in activating the M-phase form of the $CDC2$ kinase, if applicable at START, have the potential to generate "excitability" (i.e. an ability to flip from one state to another) that could well account for the apparently stochastic initiation of the cell cycle (Brooks et al. 1980; Smith and Martin 1973).

Finally, in mammalian cells, an important goal must be to determine how regulation of the $CDC2^S$ kinase is integrated with the pathways activated by growth factors. It is often assumed that growth factors and proto-oncogenes act directly on the processes controlling cell proliferation. In fact, they may do little more than stimulate a generalised increase in protein synthesis (i.e. growth in cell mass). The effects on cell division may follow indirectly because the accumulation of G_1 cyclins is tied to the overall rate of protein synthesis. Whether this view is correct remains to be seen. In any event, it is almost certain that a full understanding of how growth factors and proto-oncogenes regulate cell proliferation will depend on detailed knowledge of the role of $p34^{cdc2}$ at START.

Postscript

Since completion of this chapter in March 1991, there have been several important advances. One of the most significant is the discovery of candidate G_1 cyclins in organisms other than budding yeast, and their implication in oncogenesis and growth factor action. The human *PRAD1* gene, identified as a probable oncogene at a chromosome breakpoint in several parathyroid tumours (and possibly the same as the *bcl*-1 oncogene in B cell tumours) has significant sequence homology to the cyclin family (Motokura et al. 1991). It has been independently cloned, as cyclin D1 (CYCD1), from a human cDNA library, through its ability to rescue a triple mutant of budding yeast deficient in all three G_1 cyclins (*CLN1*, *CLN2* and *CLN3*) (Xiong et al. 1991). The mouse homologue (referred to as *CLY1*) has also been identified as a gene induced by the macrophage growth factor, CSF-1, in a macrophage cell line (Matsushime et al. 1991). Though not yet formally proven to function at START, the circumstantial evidence (synthesis during G_1, association with a $p34^{cdc2}$-like molecule, complementation of yeast *CLN* mutations, and the connection with neoplasia) is compelling. Screening of mouse genomic and cDNA libraries with a probe to *CLY1/CYCD1/PRAD1* has revealed two additional related genes (*CLY2* and *CLY3*) (Matsushime et al. 1991). The three genes are differentially expressed in different cell types, but more than one *CLY* gene can be co-expressed in the same cell type. This recalls the redundancy of *CLN* genes in budding yeast.

Possible G_1 cyclins have also been identified in fission yeast. The $puc1^+$ gene, isolated through rescue of *CLN*-deficient budding yeast, has marked similarities to the budding yeast *CLN* genes within the cyclin box (as well as to A-type cyclins), and is probably a fission yeast *CLN* homologue (Forsburg and Nurse 1991). The $cig1^+$ gene, isolated through polymerase chain reaction (PCR) cloning is more mysterious (Bueno et al. 1991). It is closely related to B-type cyclins, but lacks the N-terminal destruction box characteristic of mitotic cyclins, as do the other candidate G_1 cyclins. Gene disruption does not affect mitosis but elongates G_1, suggesting a role in G_1, but it does not complement *CLN* mutations in budding yeast. Paradoxically, over-expression leads to arrest in G_1.

In mammalian cells, phosphorylation of the retinoblastoma gene product, Rb (a tumour suppressor – see Chap. 2) is associated with entry into S phase. This phosphorylation is probably brought about, at least in part, by $p34^{cdc2}$ or a closely related kinase (Lin et al. 1991). One of the functions of Rb appears to be to repress transcription by binding a number of cellular transcription factors (Wagner and Green 1991). This binding is disrupted by phosphorylation and also by the E1A oncogene of adenovirus, which competes for the same binding site. Interestingly, one of the transcription factors targetted by Rb (known variously as E2F or DRTF1) forms a complex with cyclin A (Mudryj et al. 1991) – itself a target for E1A – and a tripartite complex containing DRTF1, Rb and cyclin A has been reported (Bandara et al. 1991). Conceivably, the cyclin A serves to direct $p34^{cdc2}$ to Rb, which it subsequently phosphorylates, so releasing the active transcription factor.

As discussed previously, although cyclin A undoubtedly interacts with $p34^{cdc2}$, in human cells it is primarily associated with a distinct but related protein of 32 kDa. This has now been shown to be the human homologue of *Xenopus* Eg1 (Tsai et al. 1991). It has 65% homology with $p34^{cdc2}$, binds to $p13^{suc1}$ and has H1 kinase activity. It has recently been renamed cdk2 (cyclin dependent kinase 2). In a most exciting development, compelling evidence has been obtained that it is cdk2 which is required for the initiation of DNA synthesis in *Xenopus* egg extracts, and not $p34^{cdc2}$ itself (Fang and Newport 1991). In yeasts $p34^{cdc2}$ is required at both G_1/S and G_2/M, but in higher eukaryotes it now seems clear that the two transitions have come under independent control.

Finally, the roles of *cdc25* and *wee1* have been clarified. Although *cdc25* was not previously thought to encode a phosphatase, homology has now been observed with a newly discovered phosphatase from vaccinia virus, which has a specificity for serine *and* tyrosine (Moreno and Nurse 1991). It is therefore probable that the *cdc25* product is directly responsible for dephosphorylating $p34^{cdc2}$, and biochemical evidence supporting this has been obtained (Kumagai and Dunphy 1991; Strausfeld et al. 1991). As for the *wee1* protein kinase, this is almost certainly responsible for introducing the inhibitory phosphorylations on Tyr_{15} and Thr_{14} of $p34^{cdc2}$, but not exclusively. Previous observations of continued Tyr_{15} phosphorylation in the absence of *wee1* function are now explained by the discovery of another protein kinase, mik1, which acts redundantly with *wee1* (Lundgren et al. 1991). Deletion of both genes is lethal, as a result of excessively premature entry into mitosis.

References

Alfa CE, Ducommun B, Beach D and Hyams J (1990) Distinct nuclear and spindle pole body populations of cyclin-cdc2 in fission yeast. Nature 347:680–682

Arion D, Meijer L, Brizuela L and Beach D (1988) cdc2 is a component of the M phase-specific histone H_1 kinase: evidence for identity with MPF. Cell 55:371–378

Bailly E, Dorée M, Nurse P and Bornens M (1989) $p34^{cdc2}$ is located in both nucleus and cytoplasm; part is centrosomally associated at G_2/M and enters vesicles at anaphase. EMBO J 8:3985–3995

Baitinger C, Alderto J, Poenie M, Schulman H and Steinhardt RA (1990) Multifunctional Ca^{++}/calmodulin-dependent protein kinase is necessary for nuclear envelope breakdown. J Cell Biol 111:1763–1773

Bandara LR, Adamczewski JP, Hunt T and La Thangue NB (1991) Cyclin A and the retinoblastoma gene product complex with a common transcription factor. Nature 352:249–251

Beach DH, Durkacz B and Nurse PM (1982) Functionally homologous cell cycle control genes in budding and fission yeast. Nature 300:706–709

Blow JJ and Laskey RA (1988) A role for the nuclear envelope in controlling DNA replication within the cell cycle. Nature 332:546–548

Blow JJ and Nurse P (1990) A cdc2-like protein is involved in the initiation of DNA replication in *Xenopus* egg extracts. Cell 62:855–862

Blow JJ, Sheehan MA, Watson JV and Laskey RA (1989) Nuclear structure and the control of DNA replication in the *Xenopus* embryo. J Cell Sci Suppl 12:183–195

Booher R and Beach DH (1986) Site-specific mutagenesis of $cdc2^+$, a cell cycle control gene of the fission yeast *Schizosaccharomyces pombe*. Mol Cell Biol 6:3523–3530

Booher RN, Alfa CE, Hyams JS and Beach DH (1989) The fission yeast cdc2/cdc13/suc1 protein kinase: regulation of catalytic activity and nuclear localization. Cell 58:485–497

Brizuela L, Draetta G and Beach D (1987) $p13^{suc1}$ acts in the fission yeast cell division cycle as a component of the $p34^{cdc2}$ protein kinase. EMBO J 6:3507–3514

Brizuela L, Draetta G and Beach D (1989) Activation of human CDC2 protein as a histone H1 kinase is associated with complex formation with the p62 subunit. Proc Natl Sci USA 86:4362–4366

Brooks RF (1985) The transition probability model: Successes, limitations and deficiencies. In: Rensing L and Jaeger NI (eds) Temporal order. Springer-Verlag, Berlin, pp 304–314

Brooks RF (1989) Mitosis at St. Andrews: Pulling the threads together. BioEssays 11:35–38

Brooks RF, Bennett DC and Smith JA (1980) Mammalian cell cycles need two random transitions. Cell 19:493–504

Bueno A, Richardson H, Reed SI and Russell P (1991) A fission yeast B-type cyclin functioning early in the cell cycle. Cell 66:149–159

Cross FR (1988) DAF-1, a mutant gene affecting size control, pheromone arrest and cell cycle kinetics in S. cerevisiae. Mol Cell Biol 8:4675–4684

Cross FR (1989) Further characterisation of a size control gene in Saccharomyces cerevisiae. J Cell Sci Suppl 12:117–127

Dabauvalle MC, Dorée M, Bravo R and Karsenti E (1988) Role of nuclear material in the early cell cycle of Xenopus embryos. Cell 52:525–533

Dorée M (1990) Control of M-phase by maturation-promoting factor. Curr Opin Cell Biol 2:269–273

Draetta G and Beach D (1988) Activation of cdc2 protein kinase during mitosis in human cells: cell cycle-dependent phosphorylation and subunit rearrangement. Cell 54:17–26

Draetta G, Brizeula L, Potashkin J and Beach D (1987) Identification of p34 and p13, human homologs of the cell cycle regulators of fission yeast encoded by $cdc2^+$ and $suc1^+$. Cell 50:319–325

Draetta G, Piwnica-Worms H, Morrison D, Druker B, Roberts T and Beach D (1988) Human cdc2 protein kinase is a major cell cycle regulated tyrosine kinase substrate. Nature 336:738–744

Draetta G, Luca F, Westendorf J, Brizuela L, Ruderman J and Beach D (1989) cdc2 protein kinase is complexed with both cyclin A and B: evidence for proteolytic inactivation of MPF. Cell 56:829–838

Dunphey WG and Newport JW (1989) Fission yeast p13 blocks mitotic activation and tyrosine dephosphorylation of the Xenopus cdc2 protein kinase. Cell 58:181–191

Dunphey WG, Brizuela L, Beach D and Newport J (1988) The Xenopus cdc2 protein is a component of MPF, a cytoplasmic regulator of mitosis. Cell 54:423–431

D'Urso G, Marraccino RL, Marshak DR and Roberts JM (1990) Cell cycle control of DNA replication by a homologue from human cells of the $p34^{cdc2}$ protein kinase. Science 250:786–791

Edgar BA and O'Farrell PH (1989) Genetic control of cell division pattern in the Drosophila embryo. Cell 57:177–187

Enoch T and Nurse P (1990) Mutation of fission yeast cell cycle control genes abolishes dependence of mitosis on DNA replication. Cell 60:665–673

Erikson E and Maller JL (1989) Biochemical characterisation of the $p34^{cdc2}$ protein kinase component of purified maturation-promoting factor from Xenopus eggs. J Biol Chem 264:19577–19582

Evans T, Rosenthal ET, Youngblom J, Distel D and Hunt T (1983) Cyclin: a protein specified by maternal mRNA in sea urchin eggs that is destroyed at each cleavage. Cell 33:389–396

Fairman MP, Prelich G, Tsurimoto T and Stillman B (1989) Replication of SV40 in vitro using proteins derived from a human cell extract. J Cell Sci Suppl 12:161–169

Fang F and Newport JW (1991) Evidence that the G1–S and G2–M transitions are controlled by different cdc2 proteins in higher eukaryotes. Cell 66:731–742

Featherstone C and Russell P (1991) Fission yeast $p107^{wee1}$ mitotic inhibitor is a tyrosine/serine kinase. Nature 349:808–811

Felix MA, Labbé JC, Dorée M, Hunt T and Karsenti E (1990) Triggering of cyclin degradation in interphase extracts of amphibian eggs by cdc2 kinase. Nature 346:379–382

Forsburg SL and Nurse P (1991) Identification of a G1-type cyclin $puc1^+$ in the fission yeast Schizosaccharomyces pombe. Nature 351:245–248

Furukawa Y, Piwnica-Worms H, Ernst TJ, Kanakura Y and Griffin JD (1990) cdc2 gene expression at the G1 to S transition in human T lymphocytes. Science 250:805–808

Gautier J, Norbury C, Lohka M, Nurse P and Maller J (1988) Purified maturation-promoting factor contains the product of a Xenopus homolog of the fission yeast cell cycle control gene $cdc2^+$. Cell 54:433–439

Gautier J, Matsukawa T, Nurse P and Maller J (1989) Dephosphorylation and activation of Xenopus $p34^{cdc2}$ protein kinase during the cell cycle. Nature 339:626–629

Gautier J, Minshull J, Lohka M, Glotzer M, Hunt T and Maller JL (1990) Cyclin is a component of maturation-promoting factor from Xenopus. Cell 60:487–494

Gerhart J, Wu M and Kirschner M (1984) Cell cycle dynamics of an M-phase-specific cytoplasmic factor in Xenopus laevis oocytes and eggs. J Cell Biol 98:1247–1255

*Giordano A, Whyte P. Harlow E, Franza BR, Beach D and Draetta G (1989) A 60 kD cdc2-associated polypeptide complexes with the EIA proteins in adenovirus-infected cells. Cell 58:981–990

---

* Note: The 60 kDa protein is shown to be cyclin A in Pines and Hunter 1990.

Glotzer M, Murray AW and Kirschner MW (1991) Cyclin is degraded by the ubiquitin pathway. Nature 349:132–138

Goebl M and Byers B (1988) Cyclin in fission yeast. Cell 54:739–740

Gould KL and Nurse P (1989) Tyrosine phosphorylation of the fission yeast cdc2$^+$ protein kinase regulates entry into M. Nature 342:39–45

Gould KL, Moreno S, Tonks NK and Nurse P (1990) Complementation of the mitotic activator, p80^{cdc25}, by a human protein-tyrosine phosphatase. Science 250:1573–1576

Hadwiger JA, Wittenberg C, Richardson HE, de Barros Lopes M and Reed SI (1989) A family of cyclin homologs that control G1 in yeast. Proc Natl Acad Sci (USA) 86:6255–6259

Hagan I, Hayles J and Nurse P (1988) Cloning and sequencing of the cyclin-related *cdc13$^+$* gene and a cytological study of its role in fission yeast mitosis. J Cell Sci 91:587–595

Hartwell LH (1974) *Saccharomyces cerevisiae* cell cycle. Bacteriol Rev 38:164–198.

Hayles J, Beach D, Durkacz B and Nurse PM (1986a) The fission yeast cell cycle control gene *cdc2*: isolation of a sequence *suc1* that suppresses *cdc2* mutant function. Mol Gen Genet 202:291–293

Hayles J, Aves S and Nurse P (1986b) *suc1$^+$* is an essential gene involved in both the cell cycle and growth in fission yeast. EMBO J 5:3173–3379

Howard A and Pelc SR (1953) Synthesis of desoxyribonucleic acid in normal and irradiated cells and its relation to chromosome breakage. Heredity Suppl 6:261–273

Hunt T (1989a) Maturation promoting factor, cyclin and the control of M-phase. Curr Opin Cell Biol 1:268–274

Hunt T (1989b) Under arrest in the cell cycle. Nature 342:483–484

Hunt T (1991) Destruction's our delight. Nature 349:100–101

Hutchison CJ, Brill D, Cox R, Gilbert J, Kill I and Ford CC (1989) DNA replication and cell cycle control in *Xenopus* egg extracts. J Cell Sci Suppl 12:197–212

Ingliss RJ, Langan TA, Mathews HR, Hardie DG and Bradbury EM (1976) Advance of mitosis by histone phosphokinase. Exp Cell Res 97:418–425

Kao JPY, Alderton JM, Tsien RY and Steinhardt RA (1990) Active involvement of Ca^{++} in mitotic progression of Swiss 3T3 fibroblasts. J Cell Biol 111:183–196

Karsenti E, Bravo R and Kirschner MW (1987) Phosphorylation changes associated with the early cell cycle in *Xenopus* eggs. Dev Biol 119:442–453

Kimelman D, Kirschner MW and Scherson T (1987) The events of the midblastula transition in *Xenopus* are regulated by changes in the cell cycle. Cell 48:399–407

Kipreos ET and Want JYJ (1990) Differentiation phosphorylation of c-abl in cell cycle determined by *cdc2* kinase and phosphatase activity. Science 248:217–220

Kirschner MW, Newport J and Gerhart J (1985) The timing of early developmental events in *Xenopus*. Trends Genet 1:41–47

Kishimoto T, Kuriyama R, Kondo H and Kanatani H (1982) Generality of the action of various maturation-promoting factors. Exp Cell Res 137:121–126

Kumagai A and Dunphy WG (1991) The cdc25 protein controls tyrosine dephosphorylation of the cdc2 protein in a cell-free system. Cell 64:903–914

Labbé J-C, Picard A, Karsenti E and Dorée M (1988a) An M-phase-specific protein kinase of *Xenopus* oocytes: partial purification and possible mechanism of its periodic activation. Dev Biol 127:157–169

Labbé J-C, Lee MG, Nurse P, Picard A and Dorée M (1988b) Activation at M-phase of a protein kinase encoded by a starfish homologue of the cell cycle control gene *cdc2$^+$*. Nature 335:251–254

Labbé J-C, Capony J-P, Caput D et al. (1989a) MPF from starfish oocytes at first meiotic metaphase is a heterodimer containing one molecule of *cdc2* and one molecule of cyclin B. EMBO J 8:3053–3058

Labbé J-C, Picard A, Peaucellier G, Cavadore J-C, Nurse P and Dorée M (1989b) Purification of MPF from starfish: identification as the H1 histone kinase p34^{cdc2} and a possible mechanism for its periodic activation. Cell 57:253–263

Langan TA, Gautier J, Lohka M et al. (1989) Mammalian growth-associated H1 histone kinase: a homologue of CDC28 protein kinases controlling mitotic entry in yeast and frog cells. Mol Cell Biol 9:3860–3868

Lee MG and Nurse P (1987) Complementation used to clone a human homologue of the fission yeast cell cycle control gene *cdc2$^+$*. Nature 327:31–35

Lee TH, Solomon MJ, Mumby MC and Kirschner MW (1991) INH, a negative regulator or MPF, is a form of protein phosphatase 2A. Cell 64:415–423

Lehner CF and O'Farrell PH (1989) Expression and function of *Drosophila* cyclin A during embryonic cell cycle progression. Cell 56:957–968

Lehner CF and O'Farrell PH (1990) The roles of *Drosophila* cyclins A and B in mitotic control. Cell 61:535–547

Lewin B (1990) Driving the cell cycle: M phase kinase, its partners and substrates. Cell 61:743–752

Lin BTY, Gruenwald S, Morla AO, Lee W-H and Wang JYJ (1991) Retinoblastoma cancer suppressor gene product is a substrate of the cell cycle regulator cdc2 kinase. EMBO J 10: 857–864

Lohka MJ, Hayes MK and Maller JL (1988) Purification of maturation-promoting factor, an intracellular regulator of early mitotic events. Proc Natl Acad Sci (USA) 85:3009–3013

Luca FC and Ruderman JV (1989) Control of programmed cyclin destruction in a cell-free system. J Cell Biol 109:1895–1909

Lundgren K, Walworth N, Booher R, Dembski M, Kirschner M and Beach D (1991) mik1 and wee1 cooperate in the inhibitory tyrosine phosphorylation of cdc2. Cell 64:1111–1122

Maller J, Wu M and Gerhart JC (1977) Changes in protein phosphorylation accompanying maturation of *Xenopus laevis* oocytes. Dev Biol 58:295–312

Masui Y and Markert CL (1971) Cytoplasmic control of nuclear behaviour during meiotic maturation of frog oocytes. J Exp Zool 177:129–146

Matsushime H, Roussel MF, Ashmun RA and Sherr CJ (1991) Colony-stimulating factor 1 regulates novel cyclins during the G1 phase of the cell cycle. Cell 65:701–713

Matthews HR and Huebner VD (1985) Nuclear protein kinases. Mol Cell Biochem 59:81–99

Meijer L, Pelech SL and Krebs EG (1987) Differential regulation of histone H1 and ribosomal S6 kinases during sea star oocyte maturation. Biochemistry 26:7968–7974

Meijer L, Arion D, Golsteyn R et al. (1989) Cyclin is a component of the sea urchin egg M-phase specific histone H1 kinase. EMBO J 8:2275–2282

Minshull J, Pines J, Golsteyn R et al. (1989a) The role of cyclin synthesis, modification and destruction in the control of cell division. J Cell Sci Suppl 12:77–97

Minshull J, Blow JJ and Hunt T (1989b) Translation of cyclin mRNA is necessary for extracts of activated *Xenopus* eggs to enter mitosis. Cell 56:947–956

Minshull J, Golsteyn R, Hill CS and Hunt T (1990) The A- and B-type cyclin associated cdc2 kinases in *Xenopus* turn on and off at different times in the cell cycle. EMBO J 9:2865–2875

Moreno S, Hayles J and Nurse P (1989) Regulation of p34^{cdc2} protein kinase during mitosis. Cell 58:361–372

Moreno S and Nurse P (1991) Clues to action of cdc25 protein. Nature 351:194

Moreno S, Nurse P and Russell P (1990) Regulation of mitosis by cyclic accumulation of p80^{cdc25} mitotic inducer in fission yeast. Nature 344:349–352

Morgan DO, Kaplan JM, Bishop JM and Varmus HE (1989) Mitosis-specific phosphorylation of p60^{c-src} by p34^{cdc2} associated protein kinase. Cell 57:775–786

Morla AO, Draetta G, Beach D and Wang JYJ (1989) Reversible tyrosine phosphorylation of cdc2: dephosphorylation accompanies activation during entry into mitosis. Cell 58:193–203

Motokura T, Bloom T, Kim HG, Jüppner H, Ruderman JV, Kronenberg HM and Arnold A (1991) A novel cyclin encoded by a *bcl1*-linked candidate oncogene. Nature 350:512–515

Mudryj M, Devoto SH, Hiebert SW, Hunter T, Pines J and Nevins JR (1991) Cell cycle regulation of the E2F transcription factor involves an interaction with cyclin A. Cell 65:1243–1253

Mullinger AM and Johnson RT (1983) Units of chromosome replication and packing. J Cell Sci 64:179–193

Murray AW and Kirschner MW (1989) Cyclin synthesis drives the early embryonic cell cycle. Nature 339:275–280

Murray AW, Solomon MJ and Kirschner MW (1989) The role of cyclin synthesis and degradation in the control of maturation promoting factor activity. Nature 339:280–286

Nash R, Tokiwa G, Anand S, Erickson K and Futcher AB (1988) The *WHI1*$^+$ gene of *Saccharomyces cerevisiae* tethers cell division to cell size and is a cyclin homolog. EMBO J 7:4335–4346

Nelkin B, Nichols C and Vogelstein B (1980) Protein factor(s) from mitotic CHO cells induce meiotic maturation in *Xenopus laevis* oocytes. FEBS Lett 109:233–238

Newport JW and Kirschner MW (1984) Regulation of the cell cycle during early *Xenopus* development. Cell 37:731–742

Nurse P (1990) Universal control mechanism regulating onset of M phase. Nature 344:503–508

Nurse P and Bissett Y (1981) Gene required in G1 for commitment to cell cycle and in G2 for control of mitosis in fission yeast. Nature 292:558–560

Nurse P, Thuriaux P and Nasmyth KA (1976) Genetic control of the divison cycle of the fission yeast *Schizosaccharomyces pombe*. Mol Gen Genet 146:167–178

Papkoff J, Verma IM and Hunter T (1982) Detection of a transforming gene product in cells transformed by Moloney murine sarcoma virus. Cell 29:417–426

Pardee AB (1974) A restriction point for the control of normal animal cell proliferation. Proc Natl Acad Sci (USA) 71:1286–1290

Paris J, Le Guellec R, Couturier A et al. (1991) Cloning by differential screening of a *Xenopus* cDNA coding for a protein highly homologous to cdc2. Proc Natl Acad Sci (USA) 88:1039–1043

Patel R, Twigg J, Crossley I, Golsteyn R and Whitaker M (1989) Calcium-induced chromatin condensation and cyclin phosphorylation during chromatin condensation cycles in ammonia-activated sea urchin eggs. J Cell Sci Suppl 12:129–144

Peter M, Nakagawa J, Dorée M, Labbé JC and Nigg EA (1990) In vitro disassembly of the nuclear lamina and M-phase specific phosphorylation of lamins by cdc2 kinase. Cell 61:591–602

Peters G (1989) Oncogenes at viral integration sites. In: Glover DM, Hames BD (eds) Oncogenes. IRL Press, Oxford, pp 23–66

Picard A, Peaucellier G, LeBouffant F, LePeuch C and Dorée M (1985) Role of protein synthesis and proteases in production and inactivation of maturation-promoting activity during meiotic maturation of starfish oocytes. Dev Biol 109:31–320

Picard A, Labbé JC, Peaucellier G, LeBouffant F, Peuch CJ and Dorée M (1987) Changes in the activity of the maturation-promoting factor are correlated with those of a major cyclic AMP and calcium independent protein kinase during the first mitotic cell cycles in the early starfish embryo. Dev Growth Differ 29:93–103

Piggott JR, Rai R and Carter BLA (1982) A bifunctional gene product involved in two phases of the yeast cell cycle. Nature 298:391–393

Pines J and Hunt T (1987) Molecular cloning and characterisation of the mRNA for cyclin from sea-urchin eggs, EMBO J 6:2987–2995

Pines J and Hunter T (1989) Isolation of a human cyclin cDNA: evidence for cyclin mRNA and protein regulation in the cell cycle and for interactions with p34^{cdc2}. Cell 58:833–846

Pines J and Hunter T (1990a) p34^{cdc2}: the S and the M kinase? N Biologist 2:389–401

Pines J and Hunter T (1990b) Human cyclin A is adenovirus E1A-associated protein p60 and behaves differently from cyclin B. Nature 346:760–763

Pondaven P, Meijer L and Beach D (1990) Activation of M-phase-specific histone H1 kinase by modification of the phosphorylation of its p34^{cdc2} and cyclin components. Genes Dev 4:9–17

Pringle JR and Hartwell LH (1982) The *Saccharomyces cerevisiae* cell cycle. In: Strathern J et al. (ed) The molecular biology of the yeast *Saccharomyces*. Cold Spring Harbor Laboratory Press, pp 97–142

Raff JW and Glover DM (1988) Nuclear and cytoplasmic mitotic cycles continue in *Drosophila* embryos in which DNA synthesis is inhibited with aphidicolin. J Cell Biol 107:2009–2019

Reed SI and Wittenberg C (1990) Mitotic role for the cdc28 protein kinase of *Saccharomyces cerevisae*. Proc Natl Acad Sci (USA) 87:5697–5701

Reed SI, Hadwiger JA, Richardson HE and Wittenberg C (1989) Analysis of the Cdc28 protein kinase complex by dosage suppression. J Cell Sci Suppl 12:29–37

Riabowol K, Draetta G, Brizuela L, Vandre D and Beach D (1989) The cdc2 kinase is a nuclear protein that is essential for mitotis in mammalian cells. Cell 57:393–401

Richardson HE, Wittenberg C, Cross F and Reed SI (1989) An essential G1 function for cyclin-like proteins in yeast. Cell 59:1127–1133

Rink TJ, Tsien RY and Warner AE (1980) Free calcium in *Xenopus* embryos measured with ion-selective microelectrodes. Nature 283:658–660

Roberts JM and D'Urso G (1989) Cellular and viral control of the initiation of DNA replication. J Cell Sci Suppl 12:171–182

Roy LM, Singh B, Gautier J, Arlingaus RB, Nordeen SK and Maller JL (1990) The cyclin B2 component of MPF is a substrate for the *c-mos*^{xe} proto-oncogene product. Cell 61:825–831

Russell P, Moreno S and Reed SI (1989) Conservation of mitotic controls in fission and budding yeasts. Cell 57:295–303

Sadhu K, Reed SI, Richardson H and Russell P (1990) Human homolog of fission yeast cdc25 mitotic inducer is predominantly expressed in G2. Proc Natl Acad Sci (USA) 87:5139–5143

Sagata N, Oskarson M, Copland T, Brumbaugh J and Vande Woude GF (1988) Function of c-*mos* proto-oncogene in meiotic maturation in *Xenopus* oocytes. Nature 335:519–525

Sagata N, Watanabe N, Vande Woude GF and Ikawa Y (1989a) The c-*mos* proto-oncogene product is responsible for meiotic arrest in vertebrate eggs. Nature 342:512–518

Sagata N, Daar I, Oskarsson M, Showalter SD and Vande Woude GF (1989b) The product of the *mos* proto-oncogene as a candidate "initiator" for oocyte maturation. Science 248:643–646

Schlegal R and Pardee A (1986) Caffeine-induced uncoupling of mitosis from the completion of DNA replication in mammalian cells. Science 232:1264–1266

Shenoy S, Choi J, Bagrodia S, Copeland TD, Maller JL and Shalloway D (1989) Purified maturation promoting factor phosphorylates pp60$^{c\text{-}src}$ at the sites phosphorylated during fibroblast mitosis. Cell 57:763–774

Simanis V and Nurse P (1986) The cell cycle control gene $cdc2^+$ of fission yeast encodes a protein kinase potentially regulated by phosphorylation. Cell 45:261–268

Smith JA (1982) The cell cycle and related concepts in cell proliferation. J Pathology 136:149–166

Smith JA and Martin L (1973) Do cells cycle? Proc Natl Acad Sci (USA) 70:1263–1267

Smith LD and Ecker RE (1971) The interaction of steroids with *Rana pipiens* oocytes in the induction of maturation. Dev Biol 25:232–247

Solomon MJ, Booher R, Kirschner M and Beach D (1988) Cyclin in fission yeast. Cell 54:738–739

Solomon MJ, Glotzer M, Lee TH, Phillippe M and Kirschner MW (1990) Cyclin activation of p34^{cdc2}. Cell 63:1013–1024

Steinhardt RA and Alderton J (1988) Intracellular free calcium rise triggers nuclear envelope breakdown in the sea urchin embryo. Nature 332:364–366

Strausfeld U, Labbé JC, Fesquet D, Cavadore JC, Picard A, Sadhu K, Russell P and Dorée M (1991) Dephosphorylation and activation of a p34^{cdc2}/cyclin B complex *in vitro* by human CDC25 protein. Nature 351:242–245

Sunkara PS, Wright DA and Rao PN (1979) Mitotic factors from mammalian cells induce germinal vesicle breakdown and chromosome condensation in amphibian oocytes. Proc Natl Acad Sci (USA) 76:2799–2802

Swenson KI, Farrell KM and Ruderman JV (1986) The clam embryo protein cyclin A induces entry into M phase and the resumption of meiosis in *Xenopus* oocytes. Cell 47:861–870

Tachibana K, Yanagashima N and Kishimoto T (1987) Preliminary characterisation of maturation-promoting factor from yeast *Saccharomyces cerevisiae*. J Cell Sci 88:273–281

Tsai L-H, Harlow E and Meyerson M (1991) Isolation of the human *cdk2* gene that encodes the cyclin A- and adenovirus E1A-associated p33 kinase. Nature 353:174–177

Twigg J, Patel R and Whitaker M (1988) Translational control of InsP$_3$-induced chromatin condensation during the early cell cycles of sea urchin embryos. Nature 332:366–369

Wagenaar EB (1983) The timing of synthesis of proteins required for mitosis in the cell cycle of the sea urchin embryo. Exp Cell Res 144:393–403

Wagner S and Green MR (1991) A transcriptional tryst. Nature 352:189–190

Wang J, Chenivesse X, Henglein B and Brechot C (1990) Hepatitis B virus integration in a cyclin A gene in heptocellular carcinoma. Nature 343:555–557

Ward G and Kirschner M (1990) Identification of cell cycle-regulated phosphorylation sites on nuclear lamin C. Cell 61:561–577

Wasserman WJ and Smith LD (1978) The cyclic behaviour of cytoplasmic factor controlling nuclear membrane breakdown. J Cell Biol 78:R15–R22

Watanabe N, Vande Woude GF, Ikawa Y and Sagata N (1989) Specific proteolysis of the c-*mos* proto-oncogene product by calpain on fertilisation of *Xenopus* eggs. Nature 342:505–511

Weinberg RA (1989) The Rb gene and the negative regulation of cell growth. Blood 74:529–532

Weinert T and Hartwell L (1989) Control of G2 delay by the RAD 9 gene of *Saccharomyces cerevisiae*. J Cell Sci Suppl 12:145–148

Weintraub H, Buscaglia M, Ferrez M et al. (1982) Mise en évidence d'une activité "MPF" chez *Saccharomyces cerevisiae*. C R Acad Sci (Paris) Ser III, 295:787–790

Westendorf JM, Swenson KI and Ruderman JV (1989) The role of the cyclin B in meiosis I. J Cell Biol 108:1431–1444

Whitaker M and Patel R (1990) Calcium and cell cycle control. Development 108:525–542

Whitfield WGF, Gonzalez C, Sanchez-Herrero E and Glover DM (1989) Transcripts of one of two *Drosophila* cyclin genes become localized in pole cells during embryogenesis. Nature 338:337–340

Wittenberg C and Reed SI (1988) Control of the yeast cell cycle is associated with assembly/disassembly of the Cdc28 protein kinase complex. Cell 54:1061–1072

Wittenberg C, Sugimoto K and Reed SI (1990) G1-specific cyclins of *S. cerevisiae*: cell cycle periodicity, regulation by mating pheromones and association with p34^{CDC28} protein kinase. Cell 62:225–237

Wu RS, Panusz HT, Hatch CL and Bonner WM (1986) Histones and their modifications. CRC Crit Rev Biochem 20:201–263

Xiong Y, Connolly T, Futcher B and Beach D (1991) Human D-type cyclin. Cell 65:691–699

2 Growth Factors, Oncogenes and Tumour Suppressor Genes

D.P. HOLLYWOOD and N.R. LEMOINE

Introduction

Control of cell proliferation involves a complex interplay between direct mechanisms linking a cell with adjacent cells and their interstitial stroma, and indirect mechanisms mediated through factors carried by organ vasculature. Disruption of the functional integrity of growth regulatory systems may lead to uncontrolled proliferation and abnormal cellular behaviour manifest in a most extreme form in the development of malignant tumours.

The control of proliferation, like other cellular responses, depends on interlinked hierarchical systems of intracellular communication. The multiple elements that comprise the regulatory pathways collectively allow fine control of cell proliferation and provide a system of checks and balances that is a barrier to disordered growth. Many components are now recognised: growth factors, growth factor receptors, membrane-associated and cytoplasmic signal transduction molecules, complex arrays of transcription factors, and tumour suppressor gene products. The system is not rigid, and a plasticity in the network allows particular components to be assigned different levels of importance depending on cellular differentiation and other conditions. There is also a capacity for transmodulation or cross-talk between signal transduction pathways that further enhances the sophistication of control (see Fig. 2.1).

Oncogenes represent variant genes with abnormal expression in time or space, or altered gene product, which is directly involved in the formation of the neoplastic phenotype. It is now recognised that many oncogene products are related to components of the signal transduction pathways involved in the control of cell proliferation and differentiation. Tumour suppressor genes represent a further class of genes whose repression, inactivation or loss results in cell transformation. The combined action of dominant oncogenes and tumour suppressor genes is thought to alter critical elements of the growth regulatory pathways, allowing abnormal cell proliferation.

Following a growth stimulus certain cell cycle parameters change, resulting in cell proliferation and the expansion of the cell population. In adult multicellular organisms, control of proliferation depends on spatial or temporal restriction of the cellular growth fraction. Thus, during embryogenesis, organ

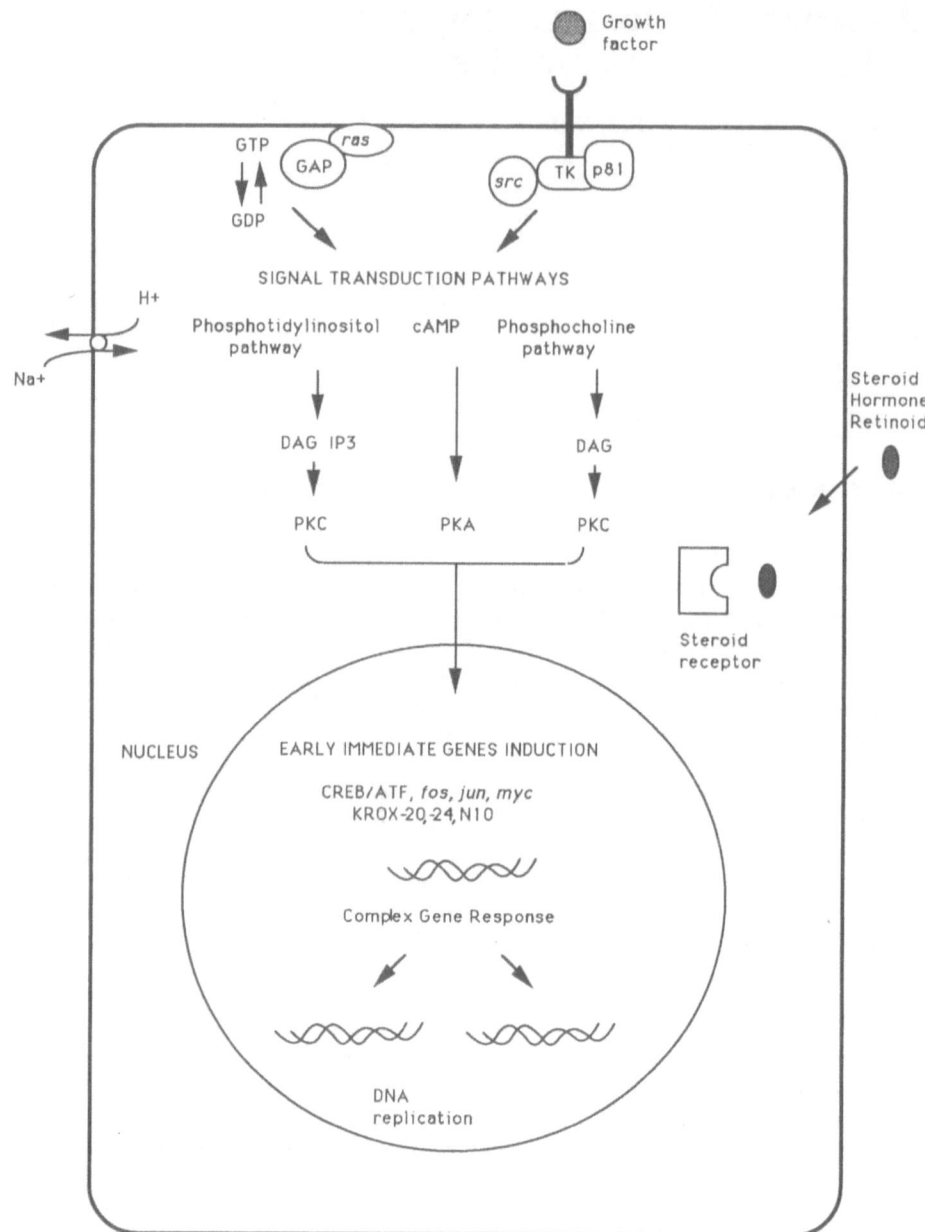

Figure 2.1. Schematic representation of the events following the presence of a growth stimulus. Production of a growth factor and interaction with the cognate receptor leads to the generation of a "signal transduction particle", the activation of specific signal transduction pathways and the induction of the early immediate gene response. Abbreviations: TK, transmembrane receptor tyrosine kinase; GAP, GTPase activating protein; p81, phosphotidylinositol-3-kinase; IP3, inositol-1,4,5-triphosphate; DAG, 1,2-diacylglycerol; PKC, protein kinase C; PKA, protein kinase A. *Italicised three-letter abbreviations* represent oncogenes.

development, wound repair and the immune response, the growth fraction increases, although control over the total cell number is retained. In contrast, the growth response of cultured cells may involve both an increase in the growth fraction and a shortened cell-cycle time, and there may be no control on total cell number.

Growth Factors and Growth Factor Receptors

Extracellular growth factors bind to their cognate receptors, resulting in the activation of one or more signal transduction pathways. The growth signal passes through sequential biochemical intermediates to target genes in the nucleus, most notably the immediate early genes c-*fos*, c-*jun*, the end-result being passage of cells from the non-proliferative quiescent state (G_0) into the replicative cell cycle (G_1, S, G_2 and M phases). The principal effect of extracellular growth factors is exercised at the transition of cells from G_0 to G_1, and progression through the remainder of the cell cycle is to a great extent independent of extracellular influences. Critical junctures within G_1 have been identified, the C, V, and R points separating competence, entry (G_{1a}), progression (G_{1b}) and assembly (G_{1c}); however, at present, there is little correlation between the G_1 subphases and specific biochemical events (Pardee 1989).

A wide variety of extracellular growth factors (GFs) are now recognised, encompassing polypeptides (epidermal, EGF; fibroblast, FGF; platelet-derived, PDGF), oligopeptides (bombesin, bradykinin, serotonin) and steroid-based compounds (oestrogens, androgens, retinoids). Autocrine, paracrine and endocrine pathways all operate to control normal proliferation and each may be subverted in tumorigenesis (Sporn 1989). For instance, autocrine growth stimulation may follow the production of bombesin and gastrin releasing peptide (GRP)-like peptides by small cell lung carcinoma cell lines (Moody et al. 1981), and gonadotrophin releasing hormone-like peptides by prostatic carcinoma cell lines (Qayum et al. 1990). Tumour cells may also produce growth factors that act in a paracrine fashion stimulating the proliferation of adjacent non-neoplastic cell populations, for example the stimulation of angiogenesis by FGF-like compounds (Gospodarowicz et al. 1987).

Most growth factor receptors belong to one of the four classes of transmembrane protein receptor with intrinsic tyrosine kinase activity. The interaction of a ligand with its transmembrane tyrosine kinase receptor results in receptor oligomerisation, activation of the cytoplasmic tyrosine kinase domain and transfer of the growth signal to the next pathway component (Carpenter 1987; Heldin and Westermark 1989). A recent model for receptor action suggests that the receptor tyrosine kinase domain may form a complex with regulatory molecules forming a "signal transfer particle" (Ullrich and Schlessinger 1989). Ligand binding alters the composition of the "signal particle", presenting an active kinase domain for phosphorylation of specific intracellular substrates. Several forms of the receptor complex may exist, depending on the identity of the regulatory proteins involved (Margolis et al. 1989).

The action of transforming growth factor β (TGFβ) will be used to exemplify the role of growth factors in cell proliferation. TGFβ possesses both stimulatory and inhibitory growth properties, exists in more than one form (TGFβ1, β2, β3, β4 and β5) and belongs to a wider family of growth-regulatory proteins including inhibin, activin and müllerian inhibitory substance. The effects exerted by TGFβ are diverse, but it is one of the most important and powerful factors involved in the control of mesenchymal and epithelial proliferation and differentiation. The growth of most epithelial cells and lymphocytes is inhibited by TGFβ, for example, subcutaneous TGFβ implants limit murine mammary epithelial growth (Silberstein and Daniel 1987). The expression of TGFβ mRNA during liver regeneration parallels the diminution of DNA synthesis, providing further support for a role in the limitation of cell proliferation. TGFβ has also been demonstrated to inhibit neoplastic proliferation, and the production of TGFβ by the MCF7 human breast carcinoma cell line provides an example of an autocrine inhibitory system (Knabbe et al. 1987). Stimulatory functions include the production of specialised extracellular matrices by cells such as osteoblasts and Schwann cells. Osteoblasts in culture produce both the latent form of TBFβ and the TGFβ receptor suggesting an autocrine mechanism in the control of bone growth (Robey et al. 1987).

Recently a model has been proposed to explain the complex phenotypic responses following TGFβ action. In the chicken chorioallantoic membrane model, the presence of a TGFβ1 concentration gradient results in a variable effect on the proliferation of different cell types, so that there is accumulation of the appropriate proportions of endothelial, epithelial and fibroblastic cells, and concomitant angiogenesis (Moses et al. 1990).

In vascular smooth muscle, low levels of TGFβ1 stimulate cell proliferation indirectly through the induction of another growth factor PDGF-AA (Battegay et al. 1990). In contrast, inhibition of proliferation at higher TGFβ1 concentrations follows the reduced expression of the PDGF receptor α-subunit. The indirect activation of PDGF and its receptor by TGFβ1 may be an example of a more widely applicable mechanism of growth factor action, and other receptor types may mediate TGFβ action in those cells that do not express PDGF or its receptor.

A possible connection has been suggested between TGFβ1, a tumour suppressor gene product (retinoblastoma protein) and the induction of the "immediate early gene" response. TGFβ1 may limit cell proliferation by inhibition of the "immediate early gene" response, in particular through limitation of *myc* protein levels. TGFβ1 reduces c-*myc* expression by inhibition of transcriptional initiation, but the precise mechanism is unknown. Activation of the TGFβ1 receptor could result in the direct production of an inhibitory transcription factor that binds to the 5′ regulatory region of c-*myc*, and the retinoblastoma protein ($p110^{Rb}$) has been implicated in this scenario. Certain DNA viral proteins (adenovirus E1A, E7, HPV16, and SV40 large T antigen) are known to bind $p110^{Rb}$ and to block TGFβ1 repression of c-*myc*, and the simplest hypothesis is that $p110^{Rb}$ is an inhibitory protein mediating TGFβ1 inhibition of c-*myc* transcription (Moses et al. 1990).

Other research suggests that $p110^{Rb}$ is not an absolute requirement for TGFβ-based growth inhibition. Certain human breast carcinoma cell lines do not express $p110^{Rb}$, as a result of Rb allelic deletion (MDA-MB-468) or gene rearrangement (MDA-MB-436), but in both cases the cells express biologically active TGFβ receptors as evidenced by TGFβ-induced inhibition of anchorage-independent cell growth (Ong et al. 1991). Thus, in this system, loss of p110^{Rb} does not abrogate TGFβ repression of cell growth, implying that other proteins with actions similar to that of p110^{Rb} must exist to effect growth inhibition.

Cytoplasmic Signal Transduction Pathways

The great diversity of growth factors and their receptors confers a high degree of target cell specificity, while the presence of common intracellular signalling pathways allows some limitation on the overall complexity of growth control. Early components of the signalling mechanisms include the alteration of transmembrane ion exchange, changes in intracellular pH and the stimulation of specific cytoplasmic signal transduction pathways.

The role of transmembrane ion exchange is controversial, but a wide range of cells exhibit rapid cytoplasmic alkalinisation and Na$^+$/H$^+$ exchange at the amiloride-sensitive Na$^+$/H$^+$ antiport following exposure to a growth stimulus. Blockade of Na$^+$ entry diminishes DNA synthesis, suggesting that Na$^+$ entry is a prerequisite for cellular proliferation. Supporters of a direct mitogenic role argue that alkalinisation and the increase in intracellular K$^+$ are central events in the growth response (Burns and Rozengurt 1984; Rozengurt 1986), while the opposing argument cites the fact that few enzymes show significant alteration in activity with the observed order of pH change (0.21 units) and K$^+$ increase. Cytoplasmic alkalinisation is seen as a secondary response to the increased metabolic rate accompanying DNA replication and cell proliferation, thereby limiting the acidotic tendency of proliferating cells (Rayter et al. 1989). Our view is that insufficient evidence exists to support either scenario with great conviction.

Three major intracellular mitogenic signal transduction pathways are recognised: substrate phosphorylation by transmembrane receptor tyrosine kinases, the phosphatidylinositol-calcium cascade, and the induction of the cAMP pathway.

Several proteins are known to be phosphorylated following activation of transmembrane tyrosine kinase receptors and probably constitute downstream elements of the signal transduction pathways. Substrates include phospholipase Cγ (PLCγ) (Miesenhelder et al. 1989), GTPase-activating protein (GAP) (Molloy et al. 1989), a phosphatidylinositol 3-kinase (p81) (Kaplan et al. 1987), and the c-*raf* proto-oncogene product (Morrison et al. 1989). Recently an interaction between membrane tyrosine kinases and the cytoplasmic *src* tyrosine kinases has been demonstrated (Kypta et al. 1990). PDGF-receptor activation is followed by phosphorylation of *src* family members (pp60^{src}, p59^{fyn}, p62^{yes}) and an increase in *src* kinase activity. Activation is paralleled by transient complex formation between the PDGF receptor, p81, and the *src* tyrosine kinase, suggesting a linkage between receptor activation, inositol phosphate metabolism and intracellular tyrosine kinase action.

Activation of the phosphoinositol pathway follows the phosphorylation of PLC$_\gamma$ at the cell membrane. PLC$_\gamma$ cleaves the membrane lipid phosphatidylinositol-4,5-biphosphate (PIP2) resulting in the production of 1,2-diacylglycerol (DAG) and inositol-1,4,5-triphosphate (IP3). IP3 acts to mobilise intracellular calcium from endoplasmic reticulum stores. The presence of DAG is short-lived and, in conjunction with the released calcium, activates key calcium-dependent growth regulatory enzymes, in particular protein kinase C (PKC) (Nishizuka 1986, 1988). The local increase of calcium is thought to activate a calcium-dependent protease which cleaves PKC into 60 and 50 kDa active fragments (Edelman et al. 1987). The signalling elements following PKC activation are presently unknown.

Activation of the cAMP cascade is mediated largely by membrane-associated G proteins with inhibitory (Gi) and stimulatory (Gs) subunits which control adenylate cyclase activity and the production of cAMP. Elevation of cAMP enhances the activity of cAMP-dependent protein kinases (PKA) resulting in the phosphorylation of specific cytoplasmic substrates. The target proteins remain unknown but they appear distinct from those phosphorylated by receptor and non-receptor tyrosine kinases. The substrates eventually result in induction of cAMP-dependent gene expression (Karin 1989). Target genes contain specific promoter elements known as cAMP-response elements (CRE), which possess binding sites for a family of related transcription factors, known variously as the CRE-binding (CREB) proteins or activating transcription factors (ATF). Using thyroid epithelial models, the induction of cAMP by thyroid-stimulating hormone (TSH) exposure or cAMP activators (cholera toxin, forskolin) results in non-neoplastic thyrocyte proliferation (Dumont et al. 1989). Thus in certain cells activation of the cAMP pathway alone may suffice as a positive growth signal. Activation of growth pathways in other systems may be more complex requiring the concomitant activation of cAMP-independent pathways, for example the coactivation of the membrane tyrosine kinases or the stimulation of the phosphatidylinositol cascade, which then converge on final common pathways including the induction of the "immediate early gene" response.

The *ras* family in man consists of three genes H-*ras* (Harvey), K-*ras* (Kirsten), and N-*ras*; each gene encodes a 21 kDa membrane associated protein (p21^{ras}) that shows limited sequence homology to the Gs subunit. Mutant *ras* is the most frequently activated dominant oncogene in human malignant disease, being present in up to 75% of certain malignant tumours such as pancreatic, bowel and thyroid carcinoma (Bos 1989). The function of p21^{ras} in non-neoplastic cells is unclear, but roles in the control of cell growth and in the maintenance of cell differentiation have been proposed. Activation of the *ras* oncoprotein by point mutation interferes with the normal inhibitory control by GAP that catalyses the return of p21^{ras} to its inactive (GDP-bound) form. Stimulation of p21^{ras}-binding to GTP results in the generation of a prolonged intracellular growth signal, although both the upstream and downstream elements of the putative pathway in most cell types remain obscure. Involvement in adenylate cyclase signalling is recognised in lower eukaryotic organisms, and interactions with phospholipase A2, phospholipase C, phosphatidylinositol metabolism, and transmembrane receptor activation have been proposed in man. Recently, it has been convincingly shown that stimulation of the CD3 receptor in T lymphocytes causes a conversion of *Ki-ras* and *N-ras* p21 to the active GTP-bound form, via suppression of GAP activity (Downward et al. 1990).

The requirements for mitogenic signal transduction are not rigid, and variations in the second messenger response exist depending on the particular ligand–receptor complex and the specific cell type. For example, following PDGF stimulation of fibroblasts calcium release is not dependent on phospholipase C-induced IP3 production, suggesting that in this setting the phosphatididylinositol pathway is not essential for the mitogenic response (Ullrich and Schlessinger 1989).

Immediate Early Genes

Following growth factor or mitogen exposure there is a transient increase in expression of a diverse group of nearly 100 immediate early genes (Almendral et al. 1988). Several oncogenes belong to this group. Induction of the early genes determines the expression of other genes, a subset of which collectively result in cell proliferation. The immediate early genes so far recognised encode various nucleoproteins, secretory molecules and components of the cytoskeleton and extracellular matrix (Table 2.1). Many of the nucleoproteins appear to function as transcription factors, (e.g. *fos*, *jun*, *KROX-20*, *KROX-24*).

The promoter and enhancer elements of immediate early genes form a target for cytoplasmic and nuclear intermediaries in the signal transduction pathways (Table 2.2). In vitro models have demonstrated the transcription of immediate early gene mRNA despite the presence of exogenous protein synthesis inhibitors, thus the regulatory factors that initially direct early gene mRNA expression are already present and probably dependent only on post-translational modification for activation.

Many of the identified immediate early genes have promoters that contain sequences known as serum response elements (SRE) (Table 2.2), with which serum response factor (SRF) can interact. There are several possible mechanisms whereby the interaction of binding factors and the consensus SRE might control gene expression (Treisman 1990). In the simplest scenario one would envisage an increase in SRF following growth factor stimulation, the resultant SRF–SRE interaction directly inducing a proportional increase in gene expression. However, as more than one factor can bind to the SRE, initiation of mRNA production may be more complicated, and SRF may recruit other factors such as p62 to the SRE in the formation of a initiation complex (Shaw et al. 1989a). Indeed, the composition and activity of the transcription factor complex may depend on the pathways activated by the afferent signal to confer some specificity in the growth response. It is now appreciated that the SRE may act as a site for competition between positive regulatory factors of varying efficacy, and that the SRE–SRF complex itself may act as an acceptor site, and further accessory proteins may act positively and negatively (Shaw et al. 1989b).

The immediate early genes all share the characteristic of rapidly inducible short half-life ("spike response") mRNA production following stimulation by serum growth factors. Most of the targets of the encoded proteins have not been identified, but they probably include linked arrays of other growth regulatory genes.

Table 2.1. Immediate early genes and their products

Gene product	Immediate early gene

1. Transcription factors
   fos family:
   jun family
   KROX family
	c-fos, fos-B, *fra*-1 v-*fos*
	c-jun, jun-B, *jun*-D v-*jun*
	KROX-20 (EGR-2), KROX-24 (EGR-1, zif/268, NGF1-A)

2. Nuclear receptor proteins	*nur/77 (N10), NGF1-B*
3. Cytokine-related proteins	*JE* *KC (N51, gro, MGSA*-melanoma growth stimulatory activity)
4. Cytoskeletal proteins	Actin Fibronectin Tropomyosin
5. Others	TF (Tissue factor) PAI (Plasminogen activator inhibitor)

Table 2.2. Transcription factors and their consensus DNA binding sites

Transcription factor family	Consensus DNA binding site
Serum response factor (SRF)	NNCC(A/T)$_6$GGNN
jun family: c-*jun, jun*-B, *jun*-D	TGAGTCA
CREB/ATF family	TGACGTCA
OCT family	ATTTGCAT
Oestrogen receptor	GGTCA(N)$_3$TGACC
Progesterone receptor Androgen receptor Glucocorticoid receptor	AGAACA(N)$_3$TGTTCT
SP1	GGGCGG

A = adenine, G = guanine, C = cytosine, T = thymine,
N = non-conserved base.
Parentheses indicate a number of bases, e.g. (N)$_6$ = 6 non-conserved
bases. OCT = octamer transcription factor, SP1 = stimulatory
protein 1.
(A/T) indicates alternative bases which may occupy that position.
A more extensive list of factors and binding sites is given by Eliyahu
et al. (1985).

The *fos* and *jun* Genes

c-*fos* is considered to have a pivotal role in the control of gene regulation, facilitating the transfer of information between the second messengers of the cytoplasmic signal transduction pathways and the numerous genes which have to be activated for a coordinated cellular response.

The *fos* family includes three cellular genes: c-*fos, fos*-B and *fra*-1 (fos-related antigen). The viral oncogene v-*fos* is a constituent of the replication defective FBJ and FBR murine sarcoma retroviruses, expression of which results in the rapid development of osteosarcoma in the host animal.

c-*fos*, *fos*-B and *fra*-1, along with the members of the *jun* family, are components of the transcription factor complex AP-1 (activator protein-1) (Curran and Franza 1988). Transcription of *fos* mRNA and production of *fos* protein rapidly follows exposure to growth factors (PDGF, EGF, NGF – nerve GF) and mitogens such as phorbol esters (e.g. TPA, 12-*o*-tetradecanoilphorbol-13-acetate). c-*fos* mRNA is detectable within 5 min, peaking at 30–60 min and diminishing after 2 h to the low basal levels characteristic of quiescent resting cells. *fos*-B mRNA displays a similiar pattern, whereas *fra*-1 mRNA expression is relatively delayed, levels peaking after 2–3 h and remaining elevated for several hours (Bravo 1990). The *fos* nucleoproteins display a similar temporal sequence of expression, although the proteins' longer half-lives (several hours) relative to that of the mRNA damps the return to baseline levels.

Three members of the *jun* family are expressed in man: c-*jun*, *jun*-B and *jun*-D. The transforming retrovirus avian sarcoma virus (ASV 17) contains an altered version of a cellular *jun* protooncogene and expression of the viral oncogene (v-*jun*) leads to the development of fibrosarcoma in host chickens. Growth-factor stimulation results in c-*jun* and *jun*-B mRNA induction from the very low levels present in quiescent cells to peak levels at 1 h, diminishing over 8 h to baseline, detectable levels of c-*jun* and absent *jun*-B mRNA (Lamph et al. 1988). Detectable levels of *jun*-D mRNA are present in resting cells and show minimal induction on serum stimulation.

c-*jun* and *jun*-B expression occurs in certain terminally differentiated cells, suggesting a role in differentiation as well as cell proliferation. Both exhibit different spatic-temporal expression patterns during early mouse embryogenesis, *jun*-B expression being limited to narrower time windows in specific tissue types (Wilkinson et al. 1989).

Members of the *fos* and *jun* families form a variety of heterodimeric transcriptional complexes capable of binding to the palindromic AP-1 site. Moreover, heterodimerisation of the *jun* protein and CREB/ATF transcription factors with potential for binding to the CRE have been demonstrated (Benbrook and Jones 1990). The exact mechanism whereby different heterodimeric combinations might have preferential affinity for the promoter elements of specific target genes remains unknown.

The *myc* Genes

Five members of the myc family are recognised in man: c-*myc*, L-*myc*, N-*myc*, R-*myc* and B-*myc*. The function of the *myc* proteins has been the focus of intensive study, and roles in the maintenance of the proliferative state and in blockade of differentiation pathways are strongly suspected (Cole 1986).

Evidence supporting a role for c-*myc* in cell proliferation comes from several quarters. Quiescent G_0 cells express very low levels of c-*myc* mRNA and protein, but both increase dramatically on entry to G_1 and during the remainder of the proliferative cell cycle. This phenomenon is seen in a number of cell types and in response to a variety of growth stimuli. Stimulation of NIH 3T3 cells with a range of growth factors (PDGF, EGF, FGF) and phaeochromocytoma PC12 cells with NGF results in the rapid induction of c-*myc* mRNA (Muller et al. 1984). Transfection of early passage rodent fibroblasts with viral promoter-linked c-*myc* results in cellular immortalisation.

Following serum stimulation of quiescent G_0 fibroblasts the level of c-*myc* protein rapidly increases from 300 molecules per cell to approximately 10 000 molecules per cell. Thereafter in continuously proliferating cells, expression diminishes to approximately ten-fold that of basal levels (Evans 1990). mRNA and protein production remain constant during continued cell proliferation, implying active and continuous expression. Following removal of the growth stimulus, levels of c-*myc* protein and mRNA rapidly diminish. Increased c-*myc* expression accompanies the G_0 to G_1 transition and the early part of S phase, but conflicting evidence exists regarding its role in S phase and DNA replication. Antisense oligonucleotides to c-*myc* block S-phase entry in human lymphocytes but anti-*myc* antibodies fail to prevent S-phase entry in fibroblasts. Studies on *Xenopus* oocytes have suggested that c-*myc* is necessary for initiation of DNA replication but not for the process of DNA elongation.

The constant production of c-*myc* throughout the cell cycle in exponentially proliferating cells suggests that c-*myc* prevents exit to the G_0 resting state from other stages of the cell cycle. The consensus is that the c-*myc* level determines the choice of a cell to cease or continue proliferation. Transient or sustained repression of c-*myc* causes rapid disappearance of the protein and acts as a signal for the cell to enter G_0, whereas sustained expression of c-*myc* allows continued cycles of proliferation.

Other Immediate Early Genes

KROX-20 and *KROX-24* encode putative zinc finger transcription factors. *KROX-20* and *KROX-24* mRNAs are rapidly induced following serum stimulation, and levels rise within a few minutes, peak at 30 min and fall to basal undetectable levels by 2 h. *KROX* proteins are detectable at 30 min and remain elevated for several hours after termination of the growth stimulatory signal. *EGR-1* (the human homologue of *KROX-24*) is located on chromosome 5 at position q23-31 which is an area frequently deleted in human acute myeloid leukaemia, raising the possibility that loss of *EGR-1* may contribute to uncontrolled cell growth in this neoplasm. The DNA-binding site recognised by the Wilms' tumour locus (*WTL*) protein is similiar to that recognised by *EGR-1*. Both the tumour suppressor gene product (*WTL* protein) and the immediate early gene product (*EGR-1*) are capable of binding to the same DNA site, and mutations observed in Wilms' tumour patients abrogate the sequence recognition (Rauscher et al. 1990). In vivo interactions between the *WTL* protein and *EGR-1* remain a matter of speculation, but competition between the proteins for the same binding site may result in a stimulatory or inhibitory growth influence, and the tissue-specific induction of the *WTL* protein may effect growth arrest and differentiation in selected cells.

An immediate early gene encoding a putative nuclear ligand binding receptor has been recognised and called variously *N10*, *nur/77* or *NGF1-B*. The kinetics of *N10* induction are similar to that of other early genes, with a short half-life mRNA peaking at 1 h and diminishing rapidly within 2 h. The predicted amino acid sequence of the 70 kDa nucleoprotein is similiar to that of the steroid hormone receptor, the thyroid hormone receptor (*erb*-A) and the retinoic acid receptor. The proposed ligand for the receptor, its DNA binding site and the target regulatory genes remain unknown.

Steroid Receptors

Members of the steroid hormone family have profound effects on cell growth and organ development, the best known being proliferation of mammary gland ductal epithelium following stimulation by oestrogenic steroids. Oestrogen has a similar proliferative effect on endometrial glandular epithelium, whilst the concomitant action of progesterone stimulates glandular differentiation, initiating the secretory phase of endometrial regeneration. Prolonged and excessive glucocorticoid exposure results in a complex array of growth abnormalities and the development of Cushing's syndrome. Retinol and retinoic acid (RA) alter the growth pattern of many cell types; for example, exposure of the F9 embryonal carcinoma cell line to RA results in primitive endodermal differentiation, exposure of the human promyelocytic HL60 cell line to RA results in an irreversible block to cell proliferation and granulocytic differentiation, and RA inhibits the proliferation of several human breast carcinoma cell lines (T47D, BT474, MDA-MB-231, ZR-75) (Fontana et al. 1990). RA also plays a role in organogenesis; in the chick embryo, for example, it provides critical positional information in the complex development of the limb bud.

Steroid hormones are recognised as aetiological factors in certain human tumours (breast adenocarcinoma, endometrial adenocarcinoma), suggesting that deregulated steroid action may contribute to carcinogenesis. There is an increased incidence of breast carcinoma associated with prolonged oestrogen exposure following early menarche and late menopause, an increased risk of endometrial carcinoma following postmenopausal oestrogen-based hormone replacement therapy, and an association of oestrogen overproduction with breast carcinoma (Klinefelter's syndrome) and endometrial carcinoma (granulosa cell tumour of the ovary). Furthermore, the steroid-receptor status of tumour specimens can have a prognostic implication; for example, oestrogen-receptor positivity in human breast carcinoma carries a beneficial prognosis when compared to tumours with lesser or undetectable levels of receptor.

A range of cytoplasmic and nuclear receptors bind specific lipophilic growth factors, exemplified by steroid hormones and retinoic acid. The resultant ligand–receptor complexes act as ligand-regulated transcription factors that recognise specific DNA binding sites (Berg 1989). The receptors share common structural features, principally a central DNA binding region and a carboxy terminal ligand-binding domain. The cognate binding sites are known as hormone responsive elements (HREs), of which three classes are presently recognised: oestrogen responsive elements (EREs), glucocorticoid responsive elements (GREs) and thyroid hormone responsive elements (TREs).

The mechanism of transcriptional activation is incompletely understood. Present models include expression dependent only on the interaction of the steroid receptor and the HRE, where the DNA binding element is referred to as a "simple HRE", and the more complex situation of a transcriptional unit formed from multiple transcription factors binding to adjacent or juxtaposed DNA binding sites creating a "composite HRE". In the composite model, the interaction between a steroid receptor and its cognate DNA binding element can influence the action of other regulatory proteins in the formation of a transcriptional complex. Binding of the steroid-receptor complex to the HRE may enhance the localisation of other binding proteins to regulatory elements;

for example, the binding of NF-1 to the GRE of the steroid-dependent MMTV promoter is accompanied by binding of the factor/factor complex referred to as F-i (Cordingley et al. 1987). The net effect of steroid-receptor binding on transcriptional activity is variable and may result in either stimulation or re-pression of gene expression. Indeed the binding of a defined steroid receptor to the same regulatory element can have a variable influence on transcription, and this appears to be dependent on the coexisting factors in the transcrip-tional complex. For example, the rat proliferin gene contains a GRE with adjacent AP-1 sites, and the effect of the glucocorticoid receptor on proliferin expression is dependent on the nature of the *fos/jun* dimer in the composite transcriptional complex: a glucocorticoid receptor/*jun* homodimer upregulates expression, whereas a receptor–*fos/jun* heterodimer downregulates expression (Diamond et al. 1990).

The immediate consequence of steroid-receptor formation is a complex constellation of activated and repressed genes that collectively initiate the growth response. It is entirely possible that the cascade of genes regulated by steroid action includes genes activated by the other afferent signal transduction pathways, but the proteins composing the effector pathways remain obscure.

Tumour Suppressor Genes

The products of the retinoblastoma and p53 "tumour suppressor genes" are believed to play important roles in the control of cell proliferation. Both pro-teins are present in all normal cells so far examined, however their presence is not mandatory for cell survival, as tumour cell lines without the gene products do exist. Thus, both the human HL60 promyelocytic cell line and the murine L12 Abelson virus-transformed B cell line do not express either p53 allele but retain the ability to proliferate.

The p53 Gene

The p53 gene encodes a 393 amino-acid nucleoprotein present at low levels in normal cells. Confusion initially followed the incorrect identification of a mutant p53 allele as the normal gene. It is now realised that the wild-type protein is non-transforming, but dominant transforming capacity is possessed by a wide variety of mutant p53 proteins. The present model of activated p53 considers the mutant gene to act as a *trans* dominant loss-of-function mutation (dominant negative effect). In this scenario, mutation of p53 results in the loss of growth regulatory functions intrinsic to the wild-type protein. In addition, the mutant form is capable of forming oligomeric complexes with wild-type p53 thereby blocking the functions of the residual normal protein.

Several lines of experimental evidence support a role for p53 in the control of proliferation. Transfection of rat embryo fibroblasts (REF cells) with mutant p53 improves both the plating efficiency and the subsequent generation of permanent cell lines. In contrast, both moderate and high expression levels of wild-type p53 have minor effects on plating efficiency. Co-transfection of Rat-1 cells with mutant p53 and activated *ras* results in an increased number of transformants and enhanced tumorigenicity in syngeneic animals (Eliyahu et al. 1985, Finlay et al. 1989). Expression of mutant p53 in Balb/c3T3 cells reduces the dependency on PDGF for continued proliferation (Gai et al. 1988). Finally, antisense p53 RNA has been shown to limit the growth of both transformed and non-transformed cell lines (Shohat et al. 1987).

Alterations in the p53 gene are implicated in a wide range of human tumours including bronchogenic carcinoma, colonic carcinoma, breast adenocarcinoma, osteosarcoma, glioma and chronic myeloid leukaemia. In the case of colorectal malignancy, 75% of tumours demonstrate deletional loss of one p53 allele. In the majority of tumours the remaining p53 allele at the 17p11.1–13.3 locus is mutant, the protein possessing amino-acid changes characteristic of trans-forming p53. In a number of patients loss of p53 accompanies the progression from adenoma to carcinoma. The acquisition of a mutant p53 gene is thought to confer a selective growth advantage despite the presence of wild type p53. Clonal expansion then creates a population of cells in which loss or mutation of the remaining allele confers the additional growth advantage synonymous with tumour progression (Fearon and Vogelstein 1990). Indeed, colorectal tumours with p53 alterations have been observed to behave more aggressively than those with wild-type p53.

A large region of p53, encompassing 30%–40% of the coding sequence (amino acids 118–273), may be the site of point mutation resulting in the acquisition of a transforming capacity. Most of the mutant transforming p53 proteins share common features including reduced binding of the SV40 large T antigen, increased binding of the heat-shock protein hsp70, and in the rodent p53 protein the loss of epitopes (amino acids 88–109) recognised by the mono-clonal antibody Pab246 (Finlay et al. 1989). This suggests that similar con-formational changes follow a wide variety of mutational events. In addition, mutant p53 has a prolonged half-life (up to 24 h) relative to that of the wild-type protein (20–30 min). The enhanced post-translational stability results in the accumulation of mutant p53 and the greatly elevated levels observed in transformed and malignant cells. It now appears that the increased amount of mutant p53 and the associated changes in protein binding are both important in producing the transformed state.

The functions of wild-type p53 remain unclear. The protein product exists in monomeric and oligomeric forms, and it may bind other cellular proteins in a manner analogous to its complex-formation with DNA virus proteins SV40 large T antigen and adenovirus E1b. A variety of experiments suggest that p53 may function as a cell cycle protein. Hydroxyurea and nocodazole limit cell cycle progression by synchronisation and restriction of cells at the G_1/S and G_2/M boundaries. Following transfection of osteosarcoma cells (Saos-2) with either wild-type p53 or mutant p53, the ability of cells to proceed to the syn-chronisation points was examined (Diller 1990). Over-expression of mutant p53 was accompanied by progression to both the G_1/S and G_2/M boundaries. In contrast, over-expression of wild-type p53 allowed progression to the G_1/S boundary but not to G_2/M. Thus over-expression of wild-type p53 appears to block progression beyond the G_1 compartment to S phase and DNA repli-cation. A possible explanation for this may follow the observation that p53 is known to inhibit complex formation between the viral oncoprotein SV40 large

T antigen and DNA polymerase α thereby limiting DNA replication. A similar action may be exerted on the normal cellular counterpart of SV40 T that helps to initiate DNA synthesis.

The recognition of an interaction between $p34^{cdc2}$ and p53 has suggested a further role for p53 in the regulation of DNA unwinding at the commencement of DNA replication. In vitro studies have demonstrated both the phosphorylation of p53 by the $p34^{cdc2}$ kinase and tight complex association of the two proteins. Complex formation may result in specificity for target substrate molecules at defined points within the cell cycle. A model of p53 action proposes the accumulation of underphosphorylated p53 during G_1, followed by $p34^{cdc2}$ kinase action and the participation of the phosphorylated form of p53 in the onset of DNA replication. Thus it is possible that the principal function of p53 is as a ubiquitous cell cycle regulator, and that in a more restricted number of cell lineages it also functions as a tumour suppressor gene.

Retinoblastoma Protein

The retinoblastoma gene encodes a 110–114 kDa nuclear phosphoprotein ($p110^{Rb}$) which is believed to function as a suppressor of cell proliferation and growth. The binding of $p110^{Rb}$ by specific DNA tumour virus proteins, for example SV40 large T, adenovirus E1A and human papilloma virus E7 protein, is thought to abrogate the growth inhibitory properties of $p110^{Rb}$, thereby facilitating the generation of tumours in the infected host organism. Loss of the functional protein is central to the development of familial and sporadic retinoblastoma; in addition, mutations of the gene at significant frequency have been detected in several types of human tumour including small cell lung carcinoma, breast adenocarcinoma and osteosarcoma (Horowitz et al. 1990).

Recent work has suggested a role in cell cycle control, in particular inhibition of the exit from G_0/G_1 which is dependent on the phosphorylation state of $p110^{Rb}$. A variety of cells, including primary human umbilical endothelial cells, peripheral T lymphocytes, a primate kidney cell line (CV-1P), and HeLa cells, show an association between the underphosphorylated form of Rb ($p110^{Rb}$) with the G_0/G_1 state and the phosphorylated form ($pp112–114^{Rb}$) with G_2/S. Following synchronisation of cell cycle progression, phosphorylation of the $p110^{Rb}$ accompanies the shift from G_1 to S commencing at the G_1/S boundary. No difference in phosphorylation status was detectable between the S and G_2 phases, and the phosphorylation status during mitosis is unknown. Phosphorylation of $p110^{Rb}$ occurs on serine and threonine residues and it is not yet clear whether the escape from G_1 involves the induction of a specific serine/threonine kinase at the G_1/S interface, or the loss of a phosphatase active during G_0/G_1 against a background of constitutive kinase activity. A high degree of homology exists between the $p110^{Rb}$-binding sites of DNA viral oncoproteins and a sequence within the *Schizosaccharomyces pombe cdc25* gene product. Extrapolating from the known interaction of *cdc25* and *cdc2* (Russell and Nurse 1986) in the control of the *S. pombe* cell cycle and the knowledge that *cdc2* is a serine/threonine kinase, it is possible that *cdc2* or an analogue is the serine/threonine kinase that generates $pp112–114^{Rb}$ (DeCaprio et al. 1989).

Conclusion

In this brief overview we have provided examples of the influence of known components of the growth regulatory pathways, and the proposed mechanisms of action have been explored. It is clear that balanced growth control is a result of a complex interplay between growth factors, signal transduction pathways, interlinked arrays of transcription factors, various tumour suppressor gene products, and molecules constituting effector pathways. Although the fine detail of growth control may be extremely varied between cell types, it appears that major components of growth control are shared. Indeed studies on the simplest eukaryotic organisms have frequently provided the initial insight, enabling recognition of the regulatory molecules and pathways in higher animals.

It is hoped that future clarification of growth regulatory mechanisms will allow the distinguishing features of normal and abnormal proliferation to be determined, and thereby allow the development of a new armamentarium of diagnostic and therapeutic agents. Already the anti-oestrogen, tamoxifen, has been demonstrated to possess a PKC inhibitory action (Weinstein 1988), and a cAMP analogue (8-Cl-cAMP) which possesses differential activation of type 1 and type 11 cAMP-dependent protein kinase (PKA) has entered pre-clinical phase 1 studies (Cho-Chung 1990). The goal of tumour specific "biological therapy" may be realised. It is also possible that strategies for the detection of the various regulatory elements in signal transduction or other mechanisms that initiate mitogenic responses may provide means for monitoring cell proliferation in clinical material, for example by the use of imunohistological methods. However, at present this possibility has not been realised.

References

Almendral JM, Sommer D, Macdonald-Bravo H, Burckhardt J, Perera J and Bravo R (1988) Complexity of the early genetic response to growth factors in mouse fibroblasts. Mol Cell Biol 8:2140–2148

Battegay EJ, Raines EW, Siefert RA, Bowen-Pope DF and Ross R (1990) TGFb induces bimodal proliferation of connective tissue cells via complex control of an autocrine PDGF loop. Cell 63:515–524

Benbrook DM and Jones NC (1990) Heterodimer formation between CREB and JUN proteins. Oncogene 5:295–305

Berg JM (1989) DNA binding specificity of steroid receptors. Cell 57:1065–1068

Bos J (1989) *ras* oncogenes in human cancer: a review. Cancer Res 49:4682–4689

Bravo R (1990) Genes induced during the G_0/G_1 transition in mouse fibroblasts. Semin Cancer Biol 1:37–46

Burns CP and Rozengurt E (1984) Extracellular Na^+ and initiation of DNA synthesis: role of intracellular pH and K^+. J Cell Biol 98:1082–1089

Carpenter G (1987) Receptors for epidermal growth factor and other polypeptide mitogens. Annu Rev Biochem 56:81–91

Cho-Chung YS (1990) Role of cyclic AMP receptor proteins in growth, differentiation, and suppression of malignancy: new approaches to therapy. Cancer Res 50:7093–7100

Cole MD (1986) The *myc* oncogene: its role in transformation and differentiation. Annu Rev Genet 20:361–386

Cordingley MG, Riegel AT and Hager GL (1987) Steroid-dependent interaction of transcription factors with the inducible promoter of mouse mammary tumour virus in vitro. Cell 48:261–270

Curran T and Franza Jr BR (1988) Fos and Jun: the AP1 connection. Cell 55:395–397

DeCaprio JA, Ludlow JW and Lynch D, et al. (1989) The product of the retinoblastoma gene has properties of a cell cycle regulatory element. Cell 58:1085–1095

Diamond MI, Miner JN, Yoshinaga SK and Yamamoto KR (1990) Transcription factor interactions: selectors of positive and negative regulation from a single DNA element. Science 249: 1266–1272

Diller L, Kassel J and Nelson CE, et al. (1990) p53 functions as a cell cycle control protein in osteosarcomas. Mol Cell Biol 11:5752–5781

Downward J, Graves JD, Warne PH, Rayter S and Cantrell DA (1990) Stimulation of p21^{ras} upon T-cell activation. Nature 346:719–723

Dumont JE, Jauniaux JC and Roger PP (1989) The cyclic AMP-mediated stimulation of cell proliferation. TIBS 114:67–70

Edelman AM, Blumenthal DK and Krebs EG (1987) Protein serine/threonine kinases. Ann Rev Biochem 56:567–613

Eliyahu D, Michalovitch D and Oren M (1985) Overproduction of p53 antigen makes established cell highly tumorigenic. Nature 316:158–160

Evans G (1990) The *myc* oncogene. In: Carney D, Sikora K (eds) Genes and cancer. John Wiley & Sons Ltd, Chichester, pp 31–43

Fearon ER and Vogelstein B (1990) A genetic model of colorectal carcinogenesis. Cell 61:759–767

Finlay CA, Hinds PW and Levine AJ (1989) The p53 proto-oncogene can act as a suppressor of transformation. Cell 57:1083–1093

Fontana JA, Miranda D and Burrows-Mezu A (1990) Retinoic acid inhibition of human breast carcinoma proliferation is accompanied by inhibition of the synthesis of a M$_r$ 39 000 protein. Cancer Res 50:1977–1982

Gai X, Rizzo MG, Lee J, Ullrich A and Beserga R (1988) Abrogation of the requirements for added growth factors in 3T3 cells constitutively expressing the p53 and the IFF-1 genes. Oncogene Res 3:377–386

Gospodarowicz D, Neufield G and Schwiegerer L (1987) Fibroblast growth factor: structure and biological properties. J Cell Phys Suppl 5:15–26

Heldin CH and Westermark B (1989) Growth factors as transforming proteins. Eur J Biochem 184:487–496

Horowitz JM, Park SH and Bogenmann E, et al. (1990) Frequent inactivation of the retinoblastoma anti-oncogene is restricted to a subset of human tumour cells. Proc Natl Acad Sci (USA) 87:2775–2779

Kaplan DR, Whitman D and Schaffhausen B, et al. (1987) Common elements in growth factor stimulation and oncogenic transformation: 85 kD phosphoprotein and phosphatidylinositol kinase activity. Cell 50:1021–1029

Karin M (1989) Complexities of gene regulation by cAMP. TIBS 5:65–67

Knabbe CME, Lippman ME and Wakefield LM, et al. (1987) Evidence that transforming factor-beta is a hormonally regulated negative growth factor in human breast cancer cells. Cell 48: 417–428

Kypta RM, Goldberg Y, Ulug ET and Courtneidge SA (1990) Association between the PDGF receptor and members of the *src* family of tyrosine kinases. Cell 62:481–492

Lamph WW, Wamsley P, Sassone-Corsi P and Vermi IM (1988) Induction of proto-oncogene JUN/AP1 by serum and TPA. Nature 334:629–634

Margolis B, Rhee SG and Felder S, et al. (1990) EGF induces phosphorylation of phospholipase C: a potential mechanism for EGF receptor signalling. Cell 57:1101–1107

Miesenhelder J, Suh PG, Rhee SG and Hunter T (1989) Phospholipase C-γ is a substrate for the PDGF and EGF receptor protein tyrosine kinases in vivo and in vitro. Cell 57:1109–1122

Molloy CJ, Bottaro DP, Fleming TP, Marshall MS, Gibbs JB and Aaronson S (1989) PDGF induction of tyrosine phosphorylation of GTPase activating protein. Nature 342:711–714

Moody TW, Pert CB, Gazdar AF, Carney DN and Minna JD (1981) High levels of intracellular bombesin characterize human small cell carcinoma cell lines. Science 214:1246

Morrison DK, Kaplan DR, Escobedo JA, Rapp UR, Roberts TM and Williams LT (1989) Direct activation of the serine-threonine kinase activity of Raf-1 through tyrosine phosphorylation by the PDGF B-receptor. Cell 58:649–657

Moses HL, Yang EY and Pietenpol JA (1990) TGF-β stimulation and inhibition of cell proliferation: new mechanistic insights. Cell 63:245–247

Muller R, Bravo R, Burckhardt J and Curran T (1984) Induction of c-*fos* gene and protein by growth factors precedes activation of c-*myc*. Nature 312:716–718

Nishizuka Y (1986) Studies and perspectives of protein kinase C. Science 233:305–311

Nishizuka Y (1988) The molecular heterogeneity of protein kinase C and its implications for cellular regulation. Nature 334:661–668

Ong G, Sikora K and Gullick WJ (1991) Inactivation of the retinoblastoma gene does not lead to loss of TGFβ receptor or loss of response to TGFβ in breast cancer cell lines. Oncogene (in press)

Pardee AB (1989) G1 events and regulation of cell proliferation. Science 246:603–608

Quayum A, Gullick WJ, Clayton RC, Sikora K and Waxman J (1990) High affinity receptors for gonadotrophin-releasing hormone are present in human prostate cancer. Br J Cancer 62:96–99

Rauscher III FJ, Morris JF, Tournay OE, Cook DM and Curran T (1990) Binding of the Wilms' tumour locus zinc finger protein to the EGR-1 consensus sequence. Science 250:1259–1262

Rayter SI, Iwata KK, Michitsch RW, Sorvillo JM, Valenzuela DM and Foulkes JG (1989) Biochemical function of oncogenes. In: Glover DM, Hames BD (eds) Oncogenes. Oxford University Press, Oxford, pp 113–189

Robey PG, Young MF and Flanders KC, et al. (1987) Osteoblasts synthesize and respond to TGF-beta in vitro. J Cell Biol 105:457–463

Rozengurt E (1986) Early signals in the mitogenic response. Science 234:161–166

Russell P and Nurse P (1986) *cdc25*[+] functions as an inducer in mitotic control of fission yeast. Cell 45:145–153

Shaw PE, Schroter H and Nordheim A (1989a) The ability of a ternary complex to form over the serum response element correlates with serum inducibility of the human c-*fos* promoter. Cell 56:563–572

Shaw PE, Frasch S and Nordheim A (1989b) Repression of c-*fos* transcription is mediated through p67[SRF] bound to the SRE. EMBO J 8:2567–2574

Shohat O, Greenberg M, Reisman D, Oren M and Rotter V (1987) Inhibition of cell growth mediated by plasmids encoding p53 anti-sense. Oncogene 1:277–283

Silberstein GB and Daniel CW (1987) Reversible inhibition of mammary gland growth by transforming growth factor β. Science 237:291–293

Sporn MB and Todaro GJ (1989) Autocrine secretion and malignant transformation of cells. N Engl J Med 303:878–880

Treisman R (1990) The SRE: a growth factor responsive transcriptional regulator. Semin Cancer Biol 1:47–58

Ullrich A and Schlessinger J (1989) Signal transduction by receptors with tyrosine kinase activity. Cell 61:203–212

Weinstein BI (1988) The origins of human cancer: molecular mechanisms of carcinogenesis and their implications for cancer prevention and treatment – Twenty-seventh G.H.A. Clowes Memorial Award Lecture. Cancer Res 48:4135–4143

Wilkinson DG, Bhatt S, Ryseck R-P and Bravo R (1989) Tissue specific expression of c-*jun* and *jun*-B during organogenesis in the mouse. Development 106:465–471

3 The Kinetic Organisation of Tissues

B. ANSARI and P.A. HALL

Introduction

It is clear that there are significant advantages in an organism being multicellular, in particular as a consequence of increased specialisation of its component parts. The cost of this lies in the considerable investment in the molecular and cellular machinery required to regulate the organisation of the cells that make up metazoa. For example, in normal tissues of metazoan organisms the number of cells with any particular phenotype is very carefully controlled (Hall and Watt 1989a; Hall 1989). In the development of an organism from a fertilised egg there are three key requirements. First, a carefully regulated increase in cell numbers coupled with, second, the differentiation of the appropriate cell types which are, thirdly, arranged in the appropriate spatial organisation. In the adult organism there remains a continuing need to carefully define cell numbers and phenotypes and to form these cells into highly ordered tissues with characteristic spatial and kinetic organisation (see for example Wright and Alison 1984). This is, in part, determined by the proliferative activity of progenitor cell populations within these tissues, but also by the number of growth-arrested cells, loss of cells into terminally differentiated populations and by programmed cell death or apoptosis. The ordered aggregates of cells that form tissues are homeostatically regulated and are capable of responding to diverse insults, with the maintenance of tissue integrity. The balance between proliferative and non-proliferative behaviour is very carefully regulated, there being genes involved in growth arrest as well as the better characterised genes involved in mitogenesis. Many aspects of pathology derive from alterations in the regulation of these processes. Consequently an understanding of regulation of cell numbers and the control of differentiation in these cells is central to our understanding of normal development, adult tissues and the pathological processes that affect them.

The purpose of this chapter is to highlight the functional, spatial and kinetic complexity of tissues, since an understanding of such factors is central to the rational application of methods for assessing cell proliferation. We shall in particular discuss the evidence for cellular heterogeneity in various tissues, and consider how regulation of the cellular subpopulations occurs.

The General Organisation of Tissues

The fertilised egg is totipotent, but during development there is progressive limitation in possible differentiative fates. That is, the potential expression of the genome becomes progressively restricted, and as development progresses the numbers and types of cell that can generate different cell lineages become limited. The mechanisms underpinning this remain only poorly understood but include regulation of transcription, in particular by sequence-specific DNA-binding proteins or transcription factors, alteration of the physical state of chromatin, or methylation of DNA. In addition, there is a wide range of post-transcriptional mechanisms regulating mRNA and its translation into protein (Alberts et al. 1989; Gilbert 1989; Hall 1991). By post-natal life the potential fates of all somatic cells are very tightly constrained. For example, under normal circumstances cells within one lineage only ever express a very restricted subset of the genome: keratinocytes do not spontaneously express a neural phenotype. Detailed analysis of cell lineages is now becoming possible in mammalian tissue using transgenic mice, chimaeric animals, heterozygotes for X-linked genes (Griffiths et al. 1988; Kirkland 1988; Ponder et al. 1985; Thompson et al. 1990) and the application of retroviral lineage markers (Price 1987; Sanes 1989). As a consequence there have been advances in the understanding of cell lineages and clonal architecture (reviewed in Hall 1991) and the regulatory mechanisms that control them (Blau and Baltimore 1991).

Despite these advances, the observations of Leblond (1964) remain central to the division of the lineages in adult tissues into three broad groups on the basis of their proliferative characteristics. Tissues may be static and show (in the adult) no proliferative activity (e.g. nerve). Alternatively, tissues may show some evidence of proliferation, which may be either continuous (as is seen in tissues such as skin, gut, bone marrow and testis) or conditional on the need to replace cells in response to some stimulus (as in, for example, liver, pancreas, prostate and salivary gland). Within any tissue there may be proliferative compartments (by convention designated P) and populations of non-proliferative cells (designated Q; Wright and Alison 1984). If one considers the cell lineages that make up these tissues as branches of a tree then the progenitor populations exist in the trunk or stem, and this analogy led Wilson (1895) to suggest the term *stem cell* for cells from which lineages derive.

Evidence for the Existence of Stem Cells

In continually renewing tissues, several lines of evidence point to the existence of functionally and kinetically distinct subpopulations of cells which may be termed stem cells (reviewed in Hall and Watt 1989a). For example, in bone marrow, transplantation of subpopulations of haemopoietic cells into irradiated mice is followed by reconstitution of marrow (Till and McCulloch 1961). The subpopulations involved can be defined, for example, by fractionation by buoyant density and by expression of particular phenotypes (Muller-Sieberg et al. 1989). The use of chromosomal markers and the application of retroviral lineage-marking techniques point to all haemopoietic lineages having a common precursor or stem cell (Abramson et al. 1977; Dick et al. 1985; Lemischka et al. 1986), although the exact details of lineage relationships remain undefined (despite impressive looking diagrams in many haematology texts). The bone marrow is spatially organised (Western and Bainton 1979), although compared with skin and gut this is less easily discerned.

In the skin, production of keratinocytes from the stem cell zone in the basal layer is followed by migration upwards and attendant terminal differentiation and finally desquamation (Potten 1981). Radiation-injury experiments point to the existence of a subpopulation of cells from which epidermis can be regenerated (Withers 1967). A number of kinetic studies indicate that this subpopulation appears to be very slowly cycling (Clausen et al. 1984) and may have a characteristic spatial architecture termed the *epidermal proliferative unit* (Potten 1981). Finally, keratinocytes can be cultured in vitro and then successfully employed in grafting experiments to repopulate epidermis in a manner analogous to bone marrow transplant experiments (Gallico et al. 1984). Recent data suggest that stem cells do not exist in interfollicular skin but in specific parts of the hair follicles (Lenoir et al. 1988; Cotsarelis et al. 1990). Similar arguments have been used to infer that corneal stem cells exist in the limbus (Cotsarelis et al. 1989).

In the gut, there is a continuing flux from the stem cell zone along the crypt to the surface where cells are lost into the lumen. Simple observations of the distribution of proliferating cells in gut epithelium indicate that the proliferative compartment is spatially constrained, at least under normal circumstances (Wright and Alison 1984). Studies of mice heterozygous for the X-linked gene glucose-6-phosphate dehydrogenase and of chimaeric rats indicate that crypts are derived from a single precursor cell (Griffiths et al. 1988; Ponder et al. 1985; Winton et al. 1988). It is now clear that all epithelial phenotypes within the gut can be derived from a common precursor, or stem, cell and that a single stem cell can maintain a crypt (Kirkland 1988; Thompson et al. 1990; Winton et al. 1988). The available data points to the presence of a small number of stem cell(s) at or near the base of crypts, whose daughters are rapidly dividing transit-amplifying cells that can go through up to four rounds of division before finally giving rise to non-dividing, terminally diffferentiated enterocytes (Potten and Loeffler 1987, 1990).

In all three tissues (bone marrow, skin and gut) there is considerable evidence for the existence of stem cells, *all of which is indirect and inferential.* While these data point to the existence of stem cell populations, these cells remain elusive and difficult to study (Hall 1989). We do not have any markers of stem cells, with the possible exception of CNS stem cells (Lendahl et al. 1990), and they cannot be reliably defined using morphological criteria (Hall and Watt 1989a; Hall 1991). Consequently, stem cells must be operationally defined as possessing the following properties (reviewed in Hall and Watt 1989a and Potten and Loeffler 1990). First, they have an unlimited capacity for self renewal (at least in the context of the organism's lifespan). Second, when a stem cell divides, the daughters have a choice, either to remain a stem cell or to differentiate. Third, there is argued to be unidirectionality of flow from relatively undifferentiated stem cells to differentiated non-stem cells. Stem cell populations have been inferred as having a number of other characteristics (Wright and Alison 1984). They are typically slowly cycling compared with their daughter cells. It is important to realise that transit-amplifying cells are the population that usually have the highest proliferative capacity and are most rapidly cycling. Stem cells are by definition clonogenic, but it should not simply be assumed that cells that are clonogenic in some model system are necessarily stem cells. Stem cells may be either pluripotent and able to give rise to a number of different phenotypes within a lineage or unipotent. Finally, it has been repeatedly shown that stem cells are relatively more radiosensitive than other cells (Wright and Alison 1984).

While this list describes the properties that stem cells may have, there may be differences in the degree to which any putative stem cell has them (Potten and Loeffler 1990).

Regulation of Stem Cells and Lineages Derived from Them

Of central importance is the idea that stem cells and the lineages that derive from them are exquisitely regulated, with cell production exactly balancing cell loss under normal circumstances (Fig. 3.1). A number of possible mechanisms have been proposed for the regulation of stem cells and their lineages, including the possibility of purely stochastic (or probabilistic) mechanisms (Johnson and Metcalf 1977; Suda et al. 1984; reviewed in Hall and Watt 1989a). *It should be recognised that simply stating that a process is stochastic does not preclude it being regulated as opposed to being a purely random process. The crutch of probability only begs the question of what is regulating the probabilities.*

There are two extreme views concerning the regulation of stem cells. First, regulation may be intrinsic and predetermined. There are examples in biology where such mechanisms do occur, for example in the regulation of mating type in yeast (*Schizosaccharomyces pombe* and *Saccharomyces cerevisiae*) or the regulation of many cell lineages in the nematode *Caenorhabditis elegans*. Cairns has suggested that non-equivalence of DNA strands may be one mechanism for regulating stem cells (Cairns 1975) but there is little experimental support for this in higher eukaryotes (Potten et al. 1978). The weight of available evidence is in favour of the alternative view that environmental regulation is key to the control of stem cells. Indeed, even in *C. elegans*, environmental factors clearly are important in determining differentiation in cell lineages. A range of different mechanisms might be in play, including cell–cell, cell–matrix interactions, and soluble growth factors: in *C. elegans* all have been implicated in regulating lineages.

Regulation by Position

Curry and Trentin (1967) proposed that the differentiation pathways taken by the daughters of haemopoietic stem cells were determined by the micro-

Transit amplifying populations

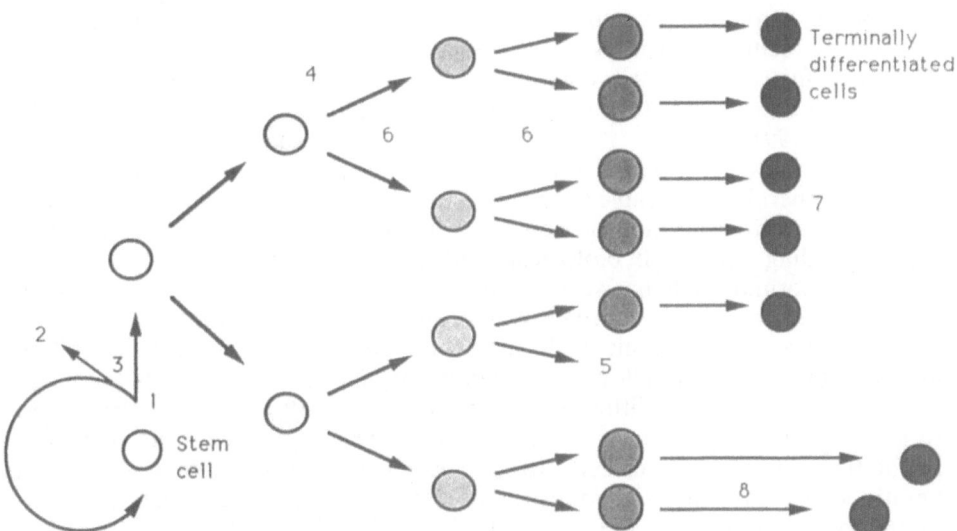

Figure 3.1. Regulation of stem cell lineages. In a stem cell hierarchy, there are a number of ways of regulating production of differentiated cells. For example, the frequency of stem cell division *(1)*, the probability of stem cell death *(2)*, the probability that daughter cells will be stem cells or committed to differentiation *(3)*, the cell cycle time of transit amplifying cells *(4)*, the probability of death of transit amplifying cells *(5)*, the number of rounds of division in the transit compartment *(6)*, and finally the lifetime of the differentiated cells *(7)*, including the duration of maturation *(8)*.

environment within the bone marrow. Based upon this idea, Schofield (1978) proposed the "niche" hypothesis, in which the environment is central to determining stem cell properties; it is certainly the case that the proliferative compartments of many tissues are normally very well defined. Recent data relating to cell adhesion, growth factors and extracellular matrix are in full accord with these views (reviewed in Hall and Watt 1989a). In particular, the idea that there are regulatory microenvironments or niches in which stem cells can exist is supported by observations relating to the adhesive properties of cellular subpopulations in model systems.

There is considerable evidence pointing to a role for adhesive mechanisms regulating stem cell behaviour in bone marrow. Using spleen colony forming assays or the ability of cells to repopulate bone marrow in irradiated mice, it has been shown that adhesive properties distinguish subpopulations of haemopoietic progenitor cells: the most adherent cells have the most stem-like properties, being resistant to 5-fluorouracil, having a high proliferative capacity, being able to reconstitute bone marrow and being able to differentiate to give all formed elements (Bearpark and Gordon 1989; Gordon et al. 1987; Gordon 1988a; Kerk et al. 1985). Furthermore, by examining human and murine marrow progenitor cells, it has been shown that the marrow stromal elements responsible for localising stem cells are highly conserved (Bearpark and Gordon 1989). The nature of the stromal factors concerned remains poorly defined, although it is clear that both cellular and extracellular factors are involved (Gordon 1988b). While the nature of the stromal cells is important

(Metcalf 1989), they need not be metabolically active (Roberts et al. 1987) and there appears to be a central role for sulphated glycosaminoglycans such as heparan sulphate, presumably in presenting growth factors to the hae-mopoietic precursors (Roberts et al. 1988; Sporn and Roberts 1988). Binding of growth factors to matrix components, in particular to proteoglycans, ap-pears to be a key regulatory mechanism in tissue organisation (Ruoslahti and Yamaguchi 1991).

While such evidence exists for bone marrow, can the ideas arising from it be more generally applicable? Evidence for the role of adhesive mechanisms in regulating differentiation processes has rapidly accumulated in the last few years, building upon cell biological studies, notably those of Bissell et al. (1982). In normal epithelia, such as gut and skin, cell proliferation is spatially restricted to specific zones, which is consistent with a role for adhesive inter-actions between clonogenic cells and microenvironments. Using a colony morphology assay that allows the identification of clones of cells with dif-ferent proliferative and differentiative properties (Hall and Watt 1989b) and simple adhesion experiments, differences in adhesive properties between sub-populations of clonogenic cells from cultures of normal human epidermal keratinocytes have been demonstrated (Hall 1990). These experiments in-dicate that cells with the greatest adhesive properties tend to have a higher proliferative capacity and lower probability of differentiation than cells with lower adhesive properties. In addition, the removal of the most adherent cells from cultures of keratinocytes depletes the population that is resistant to the differentiation-inducing effects of suspension culture. This population has been shown to correlate with the population that contains those clonogenic cells with the highest proliferative capacity (Hall and Watt 1989b). Such data suggest that there is heterogeneity of the clonogenic population of cultured human keratinocytes with respect to adhesive properties, and that the most adherent cells may be equivalent to the stem cell containing subpopulation.

These in vitro studies support the hypothesis that adhesive interactions may regulate stem cell behaviour in epithelia. It will be interesting to extend such experiments to other epithelial systems. While the molecular basis of the interactions in bone marrow appear to relate to ligand receptor interactions involving growth factors, the mechanisms in keratinocytes are uncertain but may relate to expression and functional activity of integrins or other cell–matrix and cell–cell interactions. One important consequence of the hypo-thesis that environmental factors, and in particular position, are central to the regulation of stem cells is an explanation for the difficulty in distinguishing stem cells from non-stem cells. Such a hypothesis would lead to the definition of stem cells by the possession of certain intrinsic attributes *and the position in which they exist*.

Regulation by Intercellular Communication

The behaviour of communities of cells may be regulated by the direct ex-change of low molecular weight molecules between the cytoplasm of neigh-bouring cells via intercellular junctions. That such mechanisms are important has been demonstrated in a number of systems, particularly in developmental biology. For example, in *Drosophila*, gap junction channels may be regulated and control a number of processes including pattern formation and behavioural rhythms. In this species the product of the *per* locus gene has been shown to provide a means for regulating rapid communication, amplification and synchronisation of signals involved in a range of physiological processes (Bargiello et al. 1987). Similar mechanisms have been implicated in other lower eukaryotes such as *Neurospora* (McClung et al. 1989) and *Hydra* (Fraser et al. 1987). Furthermore, antibodies that recognise a 27 kDa rat liver junction protein also recognise a gap junction antigen in *Hydra* and can inhibit cell–cell communication and consequently patterning processes (Fraser et al. 1987). Such data provide support for the notion that gap junctions are central to tissue organisation, by permitting passage of diffusible substances.

In mammalian systems there is elegant experimental data to support these ideas. Iontophoretic injection of Lucifer yellow into cultured keratinocytes allows the identification of cells coupled by gap junctions, through which the fluorescent dye can pass. Kam et al. (1987) demonstrated that terminal differentiation was associated with dramatic decline in communication, but that the proliferative compartments were coupled in groups that mimic epidermal proliferative units (Potten 1981). In the intestinal crypt, similar coupling of cells in the proliferative compartment, which is lost with differentiation, has also been reported (Bjerknes et al. 1985). The nature of the molecular signals transmitted remains uncertain, but the idea of intercellular communication and diffusible signals is concordant with ideas of positional information and spatial patterning in development and in adult tissues (Wolpert 1969).

Negative Regulators of Cell Proliferation

For 30 years a debate has continued regarding the existence of chalones, or tissue-specific negative regulators of proliferation (see Wright and Alison 1984). Although many disputed the existence of such substances, which have proved impossible to characterise, from a cybernetic point of view it is obvious that there must be inhibitory as well as stimulatory mechanisms regulating tissues, since supply and demand are so finely balanced. In recent years evidence has accrued for the existence of negative regulators, although they appear to lack the tissue specificity claimed by earlier proponents of chalones. The transforming growth factor TGFβ may also act as a negative regulator of proliferation in a number of systems (Moses et al. 1990; Roberts and Sporn 1988) including bone marrow (Dexter and White 1990). Indeed, it has been suggested that TGFβ might be considered a chalone (Parkinson and Balmain 1990). Of particular note is the ability of different cell types to respond in different ways to TGFβ, proliferation stimulated in some and inhibited in others. Moreover the same cell type may respond in different ways depending on the presence or absence of other growth factors (Roberts and Sporn 1988). The mechanism of action of TGFβ appears in part to relate to altering the phosphorylation of the retinoblastoma gene product p110 and consequently altering cell cycle control (see Chaps. 1 and 2; Laiho et al. 1990; Moses et al. 1990). In addition it is clear that TGFβ can regulate expression of extracellular matrix proteins and enzymes that may modify matrix, including metalloproteinases (Moses et al. 1990; Roberts and Sporn 1988). Given these properties, it is

of interest that the spatial distribution of TGFβ expression is very carefully
regulated both in development and in adult tissues (Roberts and Sporn 1988).

Other examples of molecules involved in negative regulation exist. Using an
in vitro colony assay it has been possible to purify to homogeneity a macrophage-
derived inhibitor of stem cell proliferation, termed SCI/MIP-1α (Graham et al.
1990). Elgjo et al. (1986) have described a pentapeptide derived from epidermis with the property of inhibiting proliferation and enhancing terminal differentiation in cultured human and murine keratinocytes. Consequently it
seems likely that while the original concept of chalones has not been fully
substantiated, there is increasing experimental evidence for the existence
of negative regulators of proliferation which may also act as stimulators of
terminal differentiation (or vice versa).

Control of Cell Number and Heterochrony

From the earlier discussion it is clear that counting mechanisms are of major
importance in biology and are central to the regulation of tissues. Both during
development and in the adult organism tissues have exquisite control of both
the type and number of their component cells, yet we know little about how
this is controlled. One simple notion is that with progressive cell divisions
there may be a progressive halving of some intracellular factor or possibly a
cell surface receptor. At some point the level may fall below a critical threshold, triggering the end of cell division. In lower eukaryotes, mutations that
alter the temporal patterns of division have been identified. These heterochronic (different time) mutations lead to cells having fates that should have
occurred earlier or later in a particular lineage. In the nematode *C. elegans*,
several such heterochronic genes have been identified (*lin-4*, *lin-14*, *lin-28* and
lin-29; see Ambros and Horwitz 1984, 1987). For example the product of
the *lin-14* gene is a nuclear protein that regulates the timing of specific post-
embryonic events (Ruvkun and Giousto 1989) such that if the *lin-14* product
level is high an early event occurs, but if low a late event occurs (Ambros
and Horwitz 1987).

Of what relevance is this to the regulation of tissues in higher eukaryotes?
It seems probable that mechanisms exist for determining numbers of divisions
and the temporal control of differentiative events in cell hierarchies (Hall and
Watt 1989a). In the gut, for example, the number of divisions that a stem
cell and its daughters undergoes within any crypt must be tightly constrained
(Potten and Loeffler 1990). One extra round of division could double the
number of cells in the crypt! It is known that the number of cells in a crypt
determines the probability of it undergoing fission and dividing into two crypts
(Cheng et al. 1986). It is worthy of note that in familial adenomatous polyposis
(FAP) the proliferative compartment is expanded and the number of bifid
(branching) crypts is dramatically increased above the normal: could it be that
the gene defect on chromosome 5 in FAP relates to the control of cell number
(i.e. the gene has a heterochronic function)? The recent identification of a
candidate gene at the FAP locus (Kinzler et al. 1991) encoding a protein with
homology to previously described muscarinic receptors possessing G-protein
activity is not inconsistent with this hypothesis. Similarly, could abnormalities
of counting mechanisms underpin diseases such as psoriasis, polycythaemia

rubra vera (increased numbers of divisions) or familial small intestinal villous atrophy (reduced numbers of divisions)?

In general we know little of the factors that regulate cell number; however in one experimental system considerable progress has been made. In the development of the rat, optic nerve neurones are ensheathed in three cell types, two morphologically distinct types of astrocyte (type 1 and type 2) and oligodendrocytes. Raff and colleagues have shown that the type-2 astrocyte and the oligodendrocyte are both derived from a common bipotential precursor, the O-2A progenitor. This cell undergoes a finite number of divisions giving rise to type-2 astrocytes before producing oligodendrocytes (Raff et al. 1985; Temple and Raff 1986). The temporal organisation of this cell lineage is tightly controlled both in vitro and in vivo by two growth factors (Hughes et al. 1988; Raff et al. 1988) elaborated by type-1 astrocytes, platelet-derived growth factor (PDGF) and ciliary neurotrophic factor (CNTF). The molecular basis of the counting or clock mechanism remains uncertain and does not appear to relate simply to growth-factor receptor levels or the target O-2A progenitor (Hart et al. 1989). Nevertheless, these important observations point to the importance of counting in biology. A recent observation of possible relevance is the progressive loss of telomeric DNA with increasing number of cell divisions, possibly constituting some form of clock (Hastie et al. 1990).

Regulation by Growth Arrest

It is evident that it is essential for organisms to negatively regulate cell proliferation. It has become clear that there are a range of genes expressed in growth-arrested cells, and that this expression is downregulated upon induction of growth by serum (Ciccarrelli et al. 1990; Manfioletti et al. 1990; Schneider et al. 1988). Several such genes were defined by use of a subtraction cDNA library enriched for mRNA sequences preferentially expressed by growth-arrested cells (Schneider et al. 1988). The nature of these genes is at present uncertain and no data are available on their spatial and temporal distribution in normal development, in adult tissues or in pathological processes. Using a similar approach, Nuell et al. (1991) have reported an evolutionarily highly conserved 30 kDa intracellular protein expressed in non-cycling cells which appears to have regulatory properties. Microinjection of mRNA encoding this protein into cycling cells rapidly induces quiescence, while treatment of quiescent cells with a 15-mer antisense oligonucleotide to the 5' sequence of prohibitin reduces protein expression and leads to proliferation (Nuell et al. 1991). Again the details of the tissue distribution of prohibitin expression remain uncertain.

Wang (1985) has reported a 57 kDa expressed by growth-arrested cells, and the gene encoding this nuclear protein has recently been cloned. Demonstration of statin using immunohistological methods reveals non-proliferating populations of cells (Wang and Kreuger 1985) and the appearance of statin immunoreactivity associated with terminal differentiation in a range of tissues (Bissonette et al. 1990). Recent observations, however, indicate that some non-proliferating cells, for example some pancreatic acinar cells, do not express detectable statin immunoreactivity (B. Ansari and P.A. Hall, unpublished) and that in established epithelial cell lines statin may be expressed in cycling cells. The significance of these observations is unclear at present. Nevertheless it seems likely that the use of antibodies to statin and to the other growth arrest associated genes may allow the characterisation of sub-populations of cells within normal and pathological tissues.

Regulation by Cell Death

Cell death can be functionally divided into *necrosis*, a process of death as a consequence of extraneous factors (e.g. ischaemia), and *apoptosis* or programmed cell death. In the context of stem cell lineages, there is evidence that apoptosis may be relevant in homeostatic control in several cell lineages, including the gut (Ijiri and Potten 1981), and is part of the process of terminal differentiation in keratinocytes (von Wangenheim 1987). Apoptosis is characterised morphologically by compaction and margination of nuclear chromatin to form sharply circumscribed masses, condensation of cytoplasm and convolution of nuclear and cellular outlines (Wyllie et al. 1984). Rapid progression of the latter culminates in nuclear fragmentation and formation of membrane-bound apoptotic bodies. These are subsequently phagocytosed and digested by nearby cells. Apoptosis is a metabolically active process (Wyllie 1980; Wyllie et al. 1984), in contrast to necrosis, and is often associated with the active synthesis of new proteins since it can be inhibited by cycloheximide. At a molecular level it is characterised by fragmentation of DNA into 200 bp fragments (or multiples thereof) corresponding to cleavage at endonuclease sensitive sites (Arends et al. 1990; Wyllie and Morris 1982). Although there is evidence of a Ca^{2+}- or Mg^{2+}-dependent endonuclease activity, the detailed nature of it and the other factors involved in apoptosis and its control are unknown. The role of signal transduction mechanisms are uncertain although Ca^{2+} fluxes appear important (McConkey et al. 1990). In addition, alterations in cell-surface protein expression probably occur, allowing phagocytes to recognise apoptotic bodies (Savill et al. 1989, 1990). In lymphoid cells the recently defined APO-1 antigen appears to be a possible trigger for programmed cell death (Trauth et al. 1989).

Apoptosis or programmed cell death occurs in a wide range of biological processes. The existence of specific genes associated with programmed cell death is supported by genetic analysis of *C. elegans* (Avery and Horwitz 1987; Ellis and Horwitz 1986; Yuan and Horwitz 1990). cDNAs associated with programmed cell death have also been defined in insect (Schwartz et al. 1990) and mammalian tissues (Leger et al. 1987). Indeed the role of regulated or programmed cell death is central to a range of processes in development (Glücksmann 1950; Saunders 1966). Assessing cell death is difficult in conventional histological material, but is of relevance to the understanding of normal and abnormal development, the maintenance of tissues and a range of disease states, and also reflects the efficacy of therapeutic modalities including radio- and chemotherapy. It is of interest that some chemotherapeutic agents may act by the induction of programmed cell death (Eastman 1990) and that the *bcl*-2 proto-oncogene is a protein found on the inner aspect of mitochondrial membranes that specifically blocks programmed cell death in B lymphocytes (Hockenbury et al. 1990).

Do Stem Cells Exist in Conditional Renewal Tissues?

If we consider some hypothetical renewing tissue, then there are two extreme possibilities with regard to regulation of cell number and differentiation. Either there is heterogeneity of proliferative and differentiative potential with the existence of subpopulations, as is seen in continually renewing tissues such as skin, gut and bone marrow and which implies the existence of stem cells. Alternatively, all the cells within this tissue are functionally equivalent with no subpopulations. Tissues such as liver, pancreas, breast and salivary gland are intermediate between continually renewing and static tissues, being composed of slowly renewing or conditionally renewal populations where there is usually little cell division but where proliferation can occur in response to certain stimuli (Leblond 1964). Such tissues are also sometimes known as expanding tissues.

Several authorities argue that in these tissues, particularly the liver, there are indeed functional subpopulations and that stem cells do exist (Reid 1990; Sell 1990). Such views are based upon a range of arguments. For example, in the fetal rat liver in vitro studies have demonstrated the existence of bipotential progenitors that can give rise to both hepatocytes and biliary radicles (Germain et al. 1988). In experiments where massive liver regeneration is induced either by the use of partial hepatectomy or by toxins and carcinogens, a precursor–product relationship is said to exist between the so-called oval cells of the rat liver and hepatocytes and biliary epithelium (Evarts et al. 1987; Germain et al. 1988; Marceau et al. 1989). The site of these progenitor cells appears to be at the junction of the liver cell plate and the biliary radicles. It should be noted, however, that in man such cells have not been identified. Moreover, while there is plausible evidence for the existence of bipotential progenitors, this is not the same as stating that stem cells exist since it does not exclude the possibility of transdifferentiation or metaplasia from one differentiated cell type to another (see Hall 1991). It is perhaps remarkable that this central question concerning the kinetic and functional composition of conditional renewal tissues remains unresolved, but this perhaps reflects the difficulties posed by our still poor understanding of stem cell biology.

The spatial and kinetic organisation of the pancreas highlights further this problem. The pancreas is composed of three defined epithelial populations: ducts, acini and islets, believed to arise during development from the primitive endoderm as dorsal and ventral outpouchings which, after numerous branchings, give rise to a complex duct system. The terminal radicles subsequently differentiate to give acini composed of cells specialised for the secretion of enzymes. The origin of the endocrine cells has been a matter of controversy. Pearse (1984) suggested that the endocrine cells of the pancreas were derived from the neural crest or from neural crest epiblast. A number of studies, in particular the use of quail–chick chimaeras, suggest that the endocrine cells are not derived from neural crest but are derived locally. In xenografts derived from the PSN-1 pancreatic carcinoma cell line, rare cells can be identified with chromogranin immunoreactivity, indicative of endocrine cell differentiation (P.A. Hall and R. Buono, unpublished). These data indicate that a cloned cell line can give rise to both exocrine (ductal) differentiation and endocrine differentiation: experimental support for the unitarian origin of all epithelial elements in the pancreas. In addition, Rao et al. (1990) have reported differentiation of pancreatic exocrine tissue to hepatocytes, but the lineage basis of this remains uncertain. Indeed there is extensive evidence for phenotypic plasticity in the pancreas derived from experiments using transgenic mice (Jhappen et al. 1990; Sandgren et al. 1991) and from in vitro experiments (P.A. Hall and N.R. Lemoine, unpublished). In particular it has become clear that interconversions between acinar and ductal phenotypes can occur. While the mechanisms required to maintain differentiated states are becoming better understood, it is evident that there are active controls regulating the stability of the differentiated state (Blau et al. 1985; Blau and Baltimore 1991; Hall 1991).

If stem cells exist in the pancreas then there should be some definable proliferative compartment spatially related to them. The pancreas is an organ in which, under normal circumstances, there is little cellular proliferation (Muller et al. 1990) but the low level of proliferation that does occur is seen in all three component epithelia. Studies to date have not allowed the characterisation of any specific proliferative compartment within the pancreas, but these have in general failed to take due consideration of the complex three-dimensional architecture of this organ, or of the need to consider the temporal relationship of slowly proliferating cells with each other. In other words, it is important to extract the time dimension as well as the third dimension from static two-dimensional histological sections. It has been suggested that all tissues undergo a process of "streaming" from some proliferative zone which contains the stem cell population, along an axis (termed a radius by Zajicek) to some final, terminally differentiated position (Wright and Alison 1984; Zajicek et al. 1985, 1987). In tissues such as gut and skin, the flux is continuous and rapid. In slowly renewing tissues such as pancreas the flux would be slow. It is reasonably easy to define the axis (radius) of migration in tissues such as gut and skin, but in complex tissues such as pancreas or salivary gland this is much more problematic.

One indirect approach to the question of the existence of stem cells in conditional renewal tissues such as the pancreas has been the use of arguments based upon the notion that neoplasia is a disorder of stem cells (see Hall 1991), but while this seems plausible there is little evidence to support it, certainly in the context of conditional renewal tissues. A number of experimental approaches to the problem exist. For example, the detailed analysis of proliferative compartments in the tissue could be performed with characterisation of the flux of cells in the normal pancreas using sequential pulse labelling experiments (Chwalinski et al. 1988; Hume and Thompson 1990). By using sequential double-labelled and computerised three-dimensional reconstruction, both spatial and temporal information might be obtained relating to proliferative compartments. By using the pulses of label separated in time one can follow their separation in space and infer both direction and rate of flux, and the origin of this flux. Another approach currently being pursued is the application of retroviral lineage markers in vivo (R. del Buono and P.A. Hall, unpublished).

A complicating factor in considering the existence of stem cells in conditional renewal populations is the possibility that the kinetic organisation of any given tissue changes during the lifespan of an organism. This notion is supported by the decreasing number of bi-potential progenitors that can be identified in cultures of rat liver with increasing post-natal age (Marceau et al.

1989). It is noteworthy that a comparable decrease in the ability to culture keratinocytes is seen with advancing age in man. A corollary of this view would be that *all* tissues have an essentially similar kinetic organisation but that the details of this will depend upon (1) the required rate of cell production (fast in skin and gut, slow in liver and pancreas), (2) the need for multi- or pluripotentiality versus a single lineage, and (3) the chronological age of the tissue (organism). A further possibility is that, under certain circumstances, differentiated (or at least non-stem) cells can take on the function of stem cells (see Hall 1991). Certainly there is increasing evidence for the ability of differentiated cells to de-differentiate and undergo further proliferation and possibly give rise to cells that pass down different lineages (see Hall 1991; Barrandon et al. 1989).

Cellular Heterogeneity in Tumours

We have seen that there is evidence for the existence of functionally and kinetically distinct populations in continually renewing tissues, although the evidence in conditional renewal tissues is less clear-cut. It must be realised that similar heterogeneity is frequently present in neoplasms. For example, there is considerable evidence that there is kinetic heterogeneity within tumours (see for example Steel 1967, 1977; Wright and Alison 1984). The basis of this relates to a number of factors. Some of the heterogeneity obviously relates to mechanical factors such as proximity to blood supply. In addition, most neoplasms recapitulate the kinetic and differentiation pathways of the tissues from which they arise (Pierce and Speers 1989; Hall 1991). Neoplasia may be associated with some loss of fidelity of differentiation control systems and the opening of new differentiation pathways and abnormal expression of the normal pathways of differentiation (Hall 1991). It has long been recognised that neoplasia may be associated with some block to differentiation pathways (Potter 1978) and this is best characterised in haemopoietic lineages (Greaves 1986). During tumour progression these blocks may become progressively more complete. In the context of lymphoma it has become apparent that much of the morphological heterogeneity seen is a consequence of the proliferative heterogeneity within the tumours (Weinberg 1989). It is important to note that there may be important kinetic differences between primary tumours and metastatic deposits.

Consequently the sampling of tumours and assessment of proliferation must take into account all these factors that might induce heterogeneity in neoplasms (Hall and Levison 1990; Hall and Woods 1990). Of particular importance must be the notion that within tumours (as in normal tissues) there are proliferative compartments and non-proliferative compartments. Therefore due consideration must be made of the appropriate populations to assess. This is by no means a new concept. Nearly 50 years ago Glücksmann investigated the morphological heterogeneity present in squamous carcinomas of the uterine cervix and reported that alterations in the proportions of different cell types in sequential biopsies could be a useful predictor of individual patient outcome (Glücksmann and Spear 1945; Glücksmann and Way 1948).

Given the availability of markers of differentiation in squamous tumours, such as involucrin and patterns of keratin expression, and the application of markers of proliferation described in this monograph, it is surprising that these observations have not been followed up.

Conclusion

There is a great need for studies to define further the spatial and temporal basis of cellular heterogeneity in normal and pathological tissues. Of particular importance is the recognition that stem cells exist in many proliferative compartments and that their regulation is central to homeostasis in those lineages. Efforts to define them and how they are regulated will be essential in order to more clearly define proliferative compartments in tissues, both normal and pathological. While in many tissues it remains uncertain as to whether sub-populations exist, *to regard normal or pathological tissues as kinetically homogenous is a recipe for disaster*!

References

Abramson S, Miller RG and Phillips RA (1977) The identification in adult bone marrow of pluri-potent and restricted stem cells of the myeloid and lymphoid systems. J Exp Med 45:1567–1579

Alberts B, Bray D, Lewis J, Raff M, Roberts K and Watson JD (eds) (1989) Molecular biology of the cell. Garland Publishing Inc., New York

Ambros V and Horwitz HR (1984) Heterochronic mutants of the nematode *Caenorhabditis elegans*. Science 266:409–416

Ambros V and Horwitz HR (1987) The *lin-14* locus of *Caenorhabditis elegans* controls the time of expression of specific post-embryonic developmental events. Genes Dev 1:398–414

Arends MJ, Morris RG and Wyllie AH (1990) Apoptosis: The role of endonuclease. Am J Pathol 136:593–608

Avery L and Horwitz HR (1987) A cell that dies during wild-type *C. elegans* development can function as a neuron in a *ced-3* mutant. Cell 51:1071–1078

Bargiello TA, Saez L, Baylies MK, Gasic G, Young MW and Spray DC (1987) The *Drosophila* clock gene *per* affects intercellular junctional communication. Nature 328:686–691

Barrandon Y, Morgan JR, Muligan RC and Green H (1989) Restoration of growth potential in paraclones of human keratinocytes by a viral oncogene. Proc Natl Acad Sci (USA) 86: 4102–4106

Bearpark AD and Gordon MY (1989) Adhesive properties distinguish subpopulations of hae-mopoietic stem cells with different spleen colony forming and marrow repopulating capacities. Bone Marrow Transplant 4:625–628

Bissell MJ, Hall GH and Parry G (1982) How does the extracellular matrix direct gene expression? J Theor Biol 99:31–68

Bissonette R, Lee M-J and Wang E (1990) The differentiation process of intestinal epithelial cells is associated with the appearance of statin, a non-proliferation specific nuclear protein. J Cell Sci 95:247

Bjerknes M, Cheng H and Erlandsen S (1985) Functional gap junctions in mouse small intestinal crypts. Anat Rec 212:364–367

Blau HM and Baltimore D (1991) Differentiation requires continuous regulation. J Cell Biol 112:781–783

Blau HM, Pavlath GK and Hardeman EC et al. (1985) Plasticity of the differentiated state. Science 230:758–766

Cairns J (1975) Mutation selection and the natural history of cancer. Nature 255:197–200

Cheng H, Bjerknes M, Amar J and Gardiner G (1986) Crypt production in normal and diseased human colonic epithelium. Anat Rec 216:44–48

Chwalinski S, Potten CS and Evans G (1988) Double labelling with bromodeoxyuridine and ^{3}H-thymidine of proliferative cells in small intestinal epithelium in the steady state and after irradiation. Cell Tissue Kinet 21:317–329

Ciccarrelli C, Philipson L and Sorrentino V (1990) Regulation of growth arrest-specific genes in mouse fibroblasts. Mol Cell Biol 10:1525–1529

Clausen OPF, Aarnaes E, Kirkhus B, Pedersen S, Thorud E and Bolund L (1984) Subpopulations of slowly cycling cells in S and G_2 phase in mouse epidermis. Cell Tissue Kinet 17:351–365

Cotsarelis G, Cheng S-Z, Dong G, Sun T-T and Lavker RM (1989) Existence of slow-cycling limbal epithelial basal cells that can be preferentially stimulated to proliferate: implications on epithelial stem cells. Cell 57:201–209

Cotsarelis G, Sun T-T and Lavker RM (1990) Label-retaining cells reside in the bulge area of pilosebaceous unit: implications for follicular stem cells, hair cycle and skin carcinogenesis. Cell 61:1329–1337

Curry JL and Trentin JJ (1967) Haemopoietic spleen colony studies. I. Growth and differentiation. Dev Biol 15:395–413

Dexter TM and White H (1990) Growth without inflation. Nature 24:380–381

Dick JE, Magli MC, Huszar D, Phillips RA and Bernstein A (1985) Introduction of a selectable gene into primitive stem cells of long-term reconstitution of the hemopoietic system of W/W^V mice. Cell 42:71–79

Eastman A (1990) Activation of programmed cell death by anticancer agents: Cisplatin as a model system. Cancer Cells 2:275–280

Elgjo K, Reichelt KL, Hennings H, Michael D and Yuspa SH (1986) Purified epidermal pentapetide inhibits proliferation and enhances terminal differentiation in cultured mouse epidermal cells. J Invest Dermatol 87:555–558

Ellis HM and Horvitz HR (1986) Genetic control of programmed cell death in the nematode *C. elegans*. Cell 44:817–829

Evarts RP, Nagy P, Marsden E and Thorgeirsson SS (1987) A precursor–product relationship exists between oval cells and hepatocytes in rat liver. Carcinogenesis 8:1737–1740

Fraser SE, Green CR, Bode HR and Gilula NB (1987) Selective disruption of gap junctional communication interferes with patterning process in Hydra. Science 237:49–55

Gallico III GG, O'Connor NE, Compton CC, Kehinde O and Green H (1984) Permanent coverage of large skin wounds with autologus cultured human epithelium. N Engl J Med 311:448–451

Germain L, Noel M, Gourdeau H and Marceau N (1988) Promotion of growth and differentiation of rat ductular oval cells in primary culture. Cancer Res 48:368–378

Gilbert SF (1989) Developmental biology, 2nd edn. Sinauer Associates Inc., Sunderland, MA, USA

Glücksmann A (1950) Cell deaths in normal vertebrate ontogeny. Biol Rev 26:59–86

Glücksmann A and Spear FG (1945) The quantitative and qualitative histological examination of biopsy material from patients treated by radiation for carcinoma of the cervix uteri. Br J Radiol 18:313–322

Glücksmann A and Way S (1948) On the choice of treatment of individual carcinomas of the cervix based on the analysis of serial biopsies. Br J Obstet Gynaecol 55:573–582

Gordon MY (1988a) Adhesive properties of haemopoietic stem cells. Br J Haematol 68:149–151

Gordon MY (1988b) Extracellular matrix of the marrow microenvironment. Br J Haematol 70:1–4

Gordon MY, Riley GP and Greaves MF (1987) Plastic adherent progenitor cells in human bone marrow. Exp Haematol 15:772–778

Graham GJ, Wright EG and Hewick R et al. (1990) Identification and characterisation of an inhibitor of haemopoietic stem cell proliferation. Nature 344:442–444

Greaves MF (1986) Differentiation linked leukaemogenesis in lymphocytes. Science 234:697–704

Griffiths DFR, Davies SJ, Williams D, Williams GT and Williams ED (1988) Demonstration of somatic mutation and colonic crypt clonality by X-linked enzyme histochemistry. Nature 333:461–463

Hall PA (1989) What are stem cells and how are they controlled? J Pathol 158:275–277

Hall PA (1990) Clonogenic human epidermal keratinocytes have differing adhesive properties which correlate with other functional properties. J Pathol 161:350a

Hall PA (1991) Differentiation, stem cells and tumour histogenesis. In: MacSween RNM and Anthony PP (eds) Recent advances in histopathology, vol 15. Churchill Livingstone, Edinburgh

Hall PA and Levison DA (1990) Assessment of cell proliferation in histological material. J Clin
    Pathol 43:184–192
Hall PA and Watt FM (1989a) Stem cells: the generation and maintenance of cellular diversity.
    Development 106:619–633
Hall PA and Watt FM (1989b) Functional characterisation of a stem cell population from cultured
    human keratinocytes. J Pathol 157:172a
Hall PA and Woods AL (1990) Immunohistological markers of cell proliferation. Cell Tissue Kinet
    23:531–549
Hart IK, Richardson WD, Heldin C-H, Westermark B and Raff MC (1989) PDGF receptors on
    cells of the oligodendrocyte-type 2 astrocyte (O-2A) cell lineage. Development 105:595–603
Hastie ND, Dempster M, Dunlop MG, Thompson AM, Green DK and Allshire RC (1990)
    Telomere reduction in human colorectal carcinoma and with aging. Nature 346:866–868
Hockenbury D, Nunez G, Milliman C, Schreiber RD and Korsmeyer SJ (1990) bcl-2 is an inner
    mitochondrial membrane protein that blocks programmed cell death. Nature 348:334–336
Hughes SM, Lillien LE, Raff MC, Rohrer H and Sendtner M (1988) Ciliary neurotrophic factor
    induces type-2 astrocyte differentiation in culture. Nature 335:70–73
Hume WJ and Thompson J (1990) Double labelling of cells with tritiated thymidine and bromo-
    deoxyuridine reveals a circadian rhythm-dependent variation in duration of DNA synthesis and
    S phase flux rates in rodent oral epithelium. Cell Tissue Kinet 23:313–323
Ijiri K and Potten CS (1981) Cell death in cell hierarchies in adult mammalian tissues. In: Potten
    CS (ed) Perspectives on mammalian cell death. Oxford University Press, Oxford
Jhappen C, Stahle C, Harkins RN, Fausto N, Smith GH and Merlino GT (1990) TGFα over-
    expression in transgenic mice induces liver neoplasia and abnormal development of the mam-
    mary gland and pancreas. Cell 61:1137–1146
Johnson GR and Metcalf M (1977) Pure and mixed erythroid colony formation in vitro stimulated
    by spleen conditioned medium with no detectable erythropoietin. Proc Natl Acad Sci (USA)
    74:3879–3882
Kam E, Watt FM and Pitts JD (1987) Patterns of junctional communication in skin: studies on
    cultured keratinocytes. Exp Cell Res 173:431–438
Kerk DK, Henry EA, Eaves AC and Eaves CJ (1985) Two classes of primitive pluripotent hae-
    mopoietic progenitor cells: separation by adherence. J Cell Physiol 125:127–134
Kinzler KW, Milbert MC and Vogelstein B et al. (1991) Identification of a gene located at
    chromosome 5q21 that is mutated in colorectal carcinomas. Science 251:1366–1370
Kirkland SC (1988) Clonal origin of columnar, mucous and endocrine cell lineages in human
    colorectal epithelium. Cancer 61:1359–1363
Laiho M, DeCaprio JA, Ludlow JW, Livingston DM and Massague J (1990) Growth inhibition by
    TGFβ linked to suppression of retinoblastoma protein phosphorylation. Cell 62:175–185
Leblond CP (1964) Classification of cell populations on the basis of their proliferative behaviour.
    J Natl Cancer Inst Monogr 14:119–148
Leger JG, Montpetit ML, Tenniswood MP (1987) Characterisation and cloning of androgen-
    repressed mRNAs from rat ventral prostate. Biochem Biophys Res Comm 147:196–203
Lemischka IR, Raulet DH and Mulligan RC (1986) Developmental potential and dynamic be-
    haviour of haematopoietic stem cells. Cell 45:917–927
Lendahl U, Zimmerman LB and McKay RDG (1990) CNS stem cells express a new class of
    intermediate filament protein. Cell 60:585–595
Lenoir M-C, Bernard BA, Pautrat G, Darmon M and Shroo B (1988) Outer root sheath cells of
    human hair follicle are able to regenerate a fully differentiated epidermis in vitro. Dev Biol
    130:610–620
Manfioletti G, Ruaro ME, Del Sal G, Philipson L and Schneider C (1990) A growth arrest-specific
    gene (gas) codes for a membrane protein. Mol Cell Biol 10:2924–2930
Marceau N, Blouin M-J, Germain L and Noel M (1989) Role of different epithelial cell types in
    liver ontogenesis, regeneration and neoplasia. In Vitro Cell Dev Biol 25:336–341
McClung CR, Fox BA and Dunlap JC (1989) The Neurospora clock gene frequency shares an
    element with the Drospophila clock gene period. Nature 339:558–562
McConkey DJ, Orrenius S and Jondal M (1990) Cellular signalling in programmed cell death
    (apoptosis). Immunol Today 11:120–121
Metcalf D (1989) The molecular control of cell division, differentiation commitment and matura-
    tion in haemopoietic cells. Nature 339:27–30
Moses HL, Yang EY and Pietenpol JA (1990) TGFβ stimulator and inhibitor of cell proliferation:
    new mechanistic insights. Cell 63:245–247

Muller-Sieberg CE, Townsend K, Weissman IL and Rennick D (1989) Proliferation and differentiation of highly enriched mouse haematopoietic stem cells and progenitor cells in response to defined growth factors. J Exp Med 167:1825–1840

Muller R, Laucke R, Trimper B and Cossel L (1990) Pancreatic cell proliferation in normal rats studied by in vivo autoradiography with ^{3}H-thymidine. Virchows Archiv 59:133–136

Nuell MJ, Stewart DA and Walker L et al. (1991) Prohibitin, an evolutionarily conserved intracellular protein that blocks DNA synthesis in normal fibroblasts and HeLa cells. Mol Cell Biol 11:1372–1381

Parkinson EK and Balmain A (1990) Chalones revisited – a possible role for transforming growth factor beta in tumour promotion. Carcinogenesis 11:195–198

Pearse AGE (1984) Islet development and the APUD concept. In: Kloppel G and Heitz PU (eds) Pancreatic pathology. Churchill Livingstone, Edinburgh, pp 125–132

Pierce GB, Speers WC (1989) Tumours as caricatures of the process of tissue renewal: prospects for therapy by directing differentiation. Cancer Res 48:1996–2004

Ponder BAJ, Schmidt GH, Wilkinson MM, Wood MJ, Monk M and Reid A (1985) Derivation of mouse intestinal crypts from single progenitor cells. Nature 313:689–691

Potten CS (1981) Cell replacement in epidermis (Keratopoiesis) via discrete units of proliferation. Int Rev Cytol 69:271–318

Potten CS and Loeffler M (1987) A comprehensive model of the crypts of the small intestine of the mouse provides insight into the mechanisms of cell migration and the proliferation hierarchy. J Theor Biol 127:381–391

Potten CS and Loeffler M (1990) Stem cells: attributes, cycles, spirals, pitfalls and uncertainties. Lessons for and from the crypt. Development 110:1001–1020

Potten CS, Hume WJ, Reid P and Cairns J (1978) The segregation of DNA in epithelial stem cells. Cell 15:899–906

Potter VR (1978) Phenotypic diversity in experimental hepatomas: the concept of partially blocked ontogeny. Br J Cancer 38:1–23

Price J (1987) Retroviruses and the study of cell lineage. Development 101:409–420

Raff MC, Abney ER and Fok-Seang J (1985) Reconstitution of a developmental clock in vitro: a critical role for astrocytes in the timing of oligodendrocyte differentiation. Cell 42:61–69

Raff MC, Lillien LE, Richardson WD, Burne JF and Noble MD (1988) Platelet derived growth factor from astrocytes drives the clock that times oligodendrocyte development in culture. Nature 333:562–565

Rao MS, Yeldandi AV and Reddy JK (1990) Differentiation and cell proliferation patterns in rat exocrine pancreas: role of type 1 and type 2 injury. Pathobiology 58:37–43

Reid LM (1990) Stem cell biology, hormone/matrix synergies and liver differentiation. Curr Opin Cell Biol 2:121–130

Roberts AB and Sporn MB (1988) Transforming growth factor β. Adv Cancer Res 51:107–145

Roberts RA, Spooncer E, Parkinson EK, Lord TI, Allen TD, Dexter TM (1987) Metabolically inactive 3T3 cells can substitute for marrow stromal cells to promote the proliferation and differentiation of multipotent haemopoietic stem cells. J Cell Physiol 132:203–214

Roberts RA, Gallagher J, Spooncer E, Allen TD, Bloomfield F and Dexter TM (1988) Heparan sulphate bound growth factors: a mechanism for stromal cell mediated haemopoiesis. Nature 332:376–378

Ruoslahti E and Yamaguchi Y (1991) Proteoglycans as modulators of growth factor activities. Cell 64:867–869

Ruvkun G and Giousto J (1989) The *Caenorhabditis elegans* heterochronic gene *lin-14* encodes a nuclear protein that forms a temporal developmental switch. Nature 338:313–320

Sandgren EP, Quaife CJ, Paulovich AG, Palmiter RD and Brinster RL (1991) Pancreatic tumour pathogenesis reflects the causative genetic lesion. Proc Natl Acad Sci (USA) 88:93–97

Sanes JR (1989) Analysing cell lineage with recombinant retrovirus. Trends Neurol Sci 12:21–28

Saunders JW (1966) Death in embryonic systems. Science 154:604–612

Savill JS, Wyllie AH, Henson JE, Walport MJ, Henson PM and Haslett C (1989) Macrophage phagocytosis of ageing neutrophils in inflammation. J Clin Invest 83:865–875

Savill J, Dransfield I, Hogg N and Haslett C (1990) Vitronectin receptor-mediated phagocytosis of cells undergoing apoptosis. Nature 343:170–173

Schneider C, King RM and Philipson L (1988) Genes specifically expressed at growth arrest in mammalian cells. Cell 54:787–793

Schofield R (1978) The relationship between the spleen-colony forming cell and the haemopoietic stem cell. Blood Cells 4:7–35

Schwartz LM, Kosz L and Kay BK (1990) Gene activation is required for developmentally programmed cell death. Proc Natl Acad Sci (USA) 87:6594–6598

Sell S (1990) Is there a liver stem cell? Cancer Res 50:3811–3815

Sporn MB and Roberts AB (1988) Peptide growth factors are multifunctional. Nature 332:217–219

Steel GG (1967) Cell loss factor in the growth of human tumours. Eur J Cancer 30:381–390

Steel GG (1977) The kinetics of tumours. Oxford University Press, Oxford

Suda T, Suda J and Ogawa M (1984) Disparate differentiation in mouse haemopoietic colonies derived from paired progenitors. Proc Natl Acad Sci (USA) 81:2520–2524

Temple S and Raff MC (1986) Clonal analysis of oligodendrocyte development in culture: evidence for a clock that counts cell divisions. Cell 44:773–779

Thompson EM, Fleming KA, Evans DJ, Fundele R, Surani MA and Wright NA (1990) Gastric endocrine cells share a clonal origin with other gut cell lineages. Development 110:477–481

Till JE and McCulloch EA (1961) A direct measurement of the radiation sensitivity of normal mouse bone marrow cells. Radiat Res 14:213–222

Trauth BC, Klas C and Peters AMJ et al. (1989) Monoclonal antibody-mediated tumour regression by induction of apoptosis. Science 245:301–304

von Wangenheim K-H (1987) Cell death through differentiation. Potential immortality of somatic cells: a failure in control of differentiation. In: Potten CS (ed) Perspectives on mammalian cell death. Oxford University Press, Oxford.

Wang E (1985) A 57 000 mol-wt protein uniquely present in non-proliferating cells and senescent human fibroblasts. J Cell Biol 100:545–559

Wang E and Kreuger JG (1985) Application of a unique monoclonal antibody as a marker for non-proliferating subpopulations of cells in some tissues. J Histochem Cytochem 33:587

Weinberg DS (1989) The role of cell-cycle activity in the generation of morphologic heterogeneity in non-Hodgkin's lymphoma. Am J Pathol 135:759–770

Western H and Bainton DF (1979) Association of alkaline phosphatase positive reticulum cells in bone marrow with granulocytic precursors. J Exp Med 150:919–937

Wilson EB (1895) The cell in development and in heredity, 1st edn. Macmillan, New York

Winton DJ, Blout MA and Ponder BAJ (1988) A clonal marker induced by mutation in mouse intestinal epithelium. Nature 333:463–466

Withers HR (1967) The dose–response relationship for irradiation of epithelial cells of mouse skin. Br J Radiol 40:187–194

Wolpert L (1969) Positional information and the spatial pattern of cellular differentiation. J Theor Biol 25:1–47

Wright NA and Alison M (1984) The biology of epithelial cell populations. Oxford University Press, Oxford

Wyllie AH (1980) Glucocorticoid induced thymocyte apoptosis is associated with endogenous endonuclease activation. Nature 284:555–556

Wyllie AH and Morris RG (1982) Hormone-induced cell death: purification and properties of thymocytes undergoing apoptosis after glucocorticoid treatment. Am J Pathol 109:78–87

Wyllie AH, Morris RG, Smith AL and Dunlop D (1984) Chromatin cleavage in apoptosis: association with condensed chromatin morphology and dependence on macromolecular synthesis. J Pathol 142:67–77

Yuan J and Horwitz HR (1990) The *Caenorhabditis elegans* genes *ced-3* and *ced-4* act cell autonomously to cause programmed cell death. Dev Biol 138:33–41

Zajicek G, Yagh C and Michaeli Y (1985) The streaming submandibular gland. Anat Rec 213: 150–158

Zajicek G, Bartfield E, Schwartz-Arad D and Michaeli Y (1987) Computerised extraction of the time dimension in histopathological sections. Appl Optics 26:3408–3412

4 Basic Methods for Assessing Cellular Proliferation

R. DOVER

Introduction

The major problem in attempting to assess proliferation is to ensure that the method used is capable of giving the data we seek. We should first consider what we want to measure. The commonest question is "can we say that cells in this tissue are being produced faster than in that?" Such questions demand a viable technique giving reproducible and unequivocal results that have discriminatory power in statistical tests, and it should be appreciated that such an ideal measurement has not yet emerged. Unfortunately, for both practical and ethical reasons, using clinical material often limits the type of procedure used. Many of the simplest methods will yield only information on the proliferative *state* of the cells not the *rate* at which they are being produced. This can produce misleading data since the question we posed asks about rates.

We must have an understanding of the basic proliferative organisation of a tissue to understand what the information that is obtained actually means (see Chap. 3). I hope to avoid lists of equations which can be found easily elsewhere (Aherne et al. 1977; Wright and Alison 1984): it is more important to be aware of some of the basic mathematical concepts in order to understand the problems and pitfalls of each technique. Only then are we equipped to know which method to employ and which to avoid when attempting to glean kinetic data from clinical or experimental material.

The Cell Cycle

The life cycle of a somatic cell between its birth at mitosis and its eventual splitting to yield new daughters at mitosis, or its demise when it dies (or in many cases differentiates and is *reproductively dead*) can be divided into a number of stages (see Fig. 4.1). The only stage apparent to the microscopist is mitosis itself. Cells produced at mitosis are believed to enter a state usually termed G_1. In most cell types G_1 is the most variable in duration and there are a number of biochemical events which occur during this time that regulate

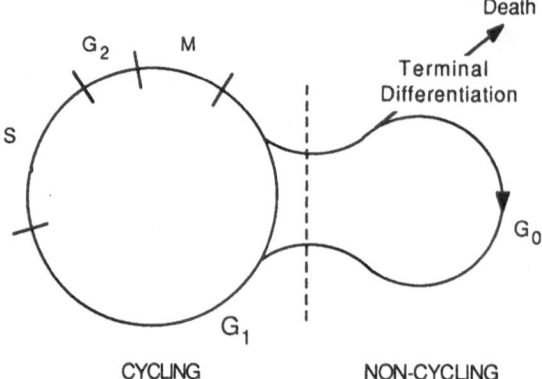

Figure 4.1. The cell cycle. The now-familiar diagram of the cell cycle denotes the phases in the life cycle of all eukaryotic cells (see Chap. 1). The period of DNA synthesis is termed S phase. The gap between the preceding mitosis and S phase is termed G_1, while the period between S phase and the following M phase is termed G_2. G_0 represents a quiescent resting state.

the duration of and exit from this phase (see Chap. 1). A separate phase, G_0, was named by Lajtha (1963) to account for cells in tissues such as liver where the bulk of the cells will not divide unless stimulated to do so (i.e. a *conditionally renewing tissue*). Such cells are not reproductively sterile by nature of their differentiated state, which is the case with, for example, a red blood cell. They are not like *static tissues*, where no proliferation occurs in the mature organ, e.g. neurones, nor are they *renewing tissues*, such as most surface epithelia and the haematopoietic system in which there is constant cell production to maintain a steady state. The conceptual point is important: is a G_0 state a normal but short part of every cycle or is it a separate state? From the point of view of our brief here, it must remain an academic point, since as there is no method currently available to separate a very long G_1 from G_0, the compartmentalisation is conceptual rather than actual.

At some point after entering G_1 the cells begin to replicate their DNA. This is termed the S phase; cells in this state can be detected by a variety of methods described elsewhere. This phase has a duration of the order of 6–16 h. After S phase the cell now has double its original DNA content and enters a second gap phase – G_2. This phase typically lasts 4–8 h, and some of the regulatory events that occur here are described in Chap. 1. After G_2 the cells enter mitosis. The time taken to move from any given point to the same point in the next cycle is the cell cycle time, T_c. The population as a whole will have a spread of T_c values. The mechanisms involved in regulating the cell cycle have been reviewed in Chap. 1.

Growth Fraction

Now we know what individual cells do, what about populations of cells? In most tissues not all the cells are dividing at any one time, in fact dividing cells are often only a small proportion. Some of the cells will be non-proliferating:

the ratio of proliferating to non-proliferating has been termed the *growth fraction*. This is sometimes denoted by the term I_p (index of proliferation):

$$I_p = \frac{P}{P + Q},$$

where P is the proportion of proliferating cells in a population and Q is the fraction of non-proliferating cells (i.e. $P + Q = 1$).

The growth fraction is important when considering proliferative rates. If the growth fraction is unity (all cells proliferating), the time taken for the population to double its size would be equal to the cell cycle time. The *population doubling time*, T_d, is a useful parameter as it gives a measurement which is simple to comprehend and to compare between two populations and, as its name suggests, is the time taken to increase the size of the population twofold. Its measurement is not so simple. If the growth fraction is less than 1, as is usual in tumours, then the T_d will be longer than T_c. It is important to note that in nearly all cases the T_c is not equal to the T_d (the exception being where the growth fraction is unity).

In most normal tissues we are interested in studying replacement of cells to maintain a steady state rather than a doubling in cell number. The time taken to replace all the cells in the population, or the time for the tissue to completely "turnover" has been called, not surprisingly, the *turnover time* T_t.

The doubling time is given, for an exponentially growing population by

$$T_d = \frac{\ln 2}{K_G},$$

where K_G is the growth rate of the population. If there is no loss K_G will equal K_B, the birth rate (see below).

The term *potential doubling time* was introduced by Steel (1968) and it is denoted by T_{pot}, sometimes by T_{pd}. This represents the time for the population to double without cell loss. T_{pot} equals T_c where there is no cell loss and the growth fraction, I_p, equals one. However T_{pot} often exceeds T_c for these very reasons.

Cell Loss

The net cell production of a population is given by

$$K_G = K_B - K_L,$$

where K_G is the growth rate for the population, K_B is the birth rate and K_L is the loss rate. We know that for the steady state

$$\frac{K_L}{K_B} = 1,$$

as the rate of production equals the rate of loss. This ratio is called ϕ, the cell loss factor, which can be calculated from

$$\phi = 1 - \left(\frac{T_{pot}}{T_d}\right).$$

Growth Curves

The measurement of doubling time can be difficult and the measurement has been equated to the measurement of *volume doubling time* obtained from a growth curve. Others have used the DNA content or even the diameter of a tumour to measure growth rate. The problems with this latter method include the involvement of vascular and connective tissue elements and a wide array of infiltrating cells from the immune system in the measurement. In tumours the problem is compounded by cell death, where cyst formation and a necrotic centre would obviously make a nonsense of some of the measurements discussed above. Few workers have attempted to correct for these errors, and most continue to equate volume to cell number directly. Some measure of how accurate this assumption is should be made in any given system.

In most cases it will be a tumour population that is being studied and the growth will be, at least in part, exponential. The data are thus normally plotted in semi-log form – the logarithm of cell number is plotted against time on an arithmetic scale. The curves produced are usually sigmoid, with an initial lag phase followed by a near linear rise and finally a plateau. One way to measure the doubling time is to extract it manually by drawing a tangent to the line. This can be unsatisfactory as a general method, as there can often be many ways to draw the same line. This is usually solved by using mathematical models. Often growth curves are fitted using what is called the Gompertz equation, one form of which is

$$V_t = V_0 \exp \left[\frac{A}{\alpha(1 - e^{-\alpha t})} \right],$$

where V is the volume at time t and 0, α is a constant and A is the initial specific growth rate. The fit is usually by computer, but the equation is not always a good fit to real data: it often bends too early at the start and too late at the plateau end.

The number of population doublings (N) during a given period can be obtained from

$$N = \left(\frac{A}{\alpha} \right) \ln 2.$$

So with some biological and mathematical provisos, T_d can be estimated for tumour populations.

Age Distribution

One might expect to be able to calculate the duration of some cycle phase from equations such as

$$MI = I_M = \frac{T_M}{T_c},$$

which simply states that the proportion of mitotic cells (mitotic index, MI) will equal the proportion of time taken for mitosis compared with the time for a whole cycle.

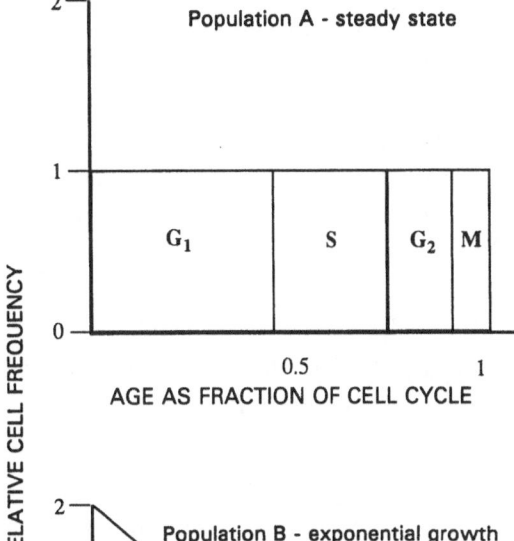

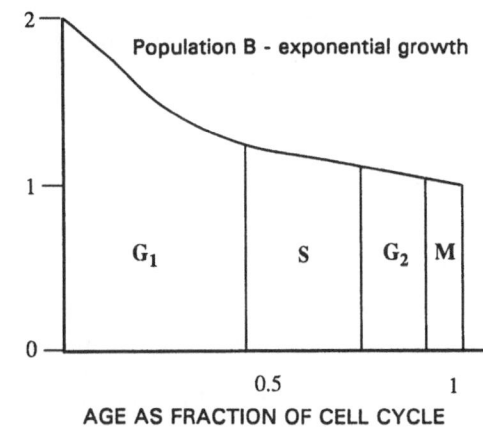

Figure 4.2. Age distributions for two populations. *Population A* is in steady state, with equal loss from all phases of the cycle. *Population B* is at the other extreme, every cell in the tissue being in exponential growth so that young cells predominate in the population. In real life, we might expect to find populations at stages in between these extremes. In some cases the loss of cells may be specific to a cycle phase, giving a more precipitous drop at some point.

Similarly one can state

$$LI = I_S = \frac{T_S}{T_c}.$$

Thus the number of cells observed in S phase should be proportional to the time spent in that state.

These approaches sound reasonable and appear to be a matter of common sense. In some cases they can actually be true, but another factor plays a part, *the age distribution*. In an exponentially growing population, such as some stages of tumour growth, every mitosis yields two G_1 cells which are retained (with no cell loss). The proportion of G_1 cells in the population will therefore be double that of mitotic cells. The population will be skewed to have a pre-ponderance of "younger" cells within its age distribution. This is shown dia-grammatically in Fig. 4.2. Taking the other extreme case, a tissue with an overall steady state, where there is random loss from each cell cycle phase, a flat-surfaced, rectangular age distribution is more accurate and our initial equations will be true if the growth fraction is unity. This situation is rare in the real world. There are methods to calculate cycle parameters from non-rectangular age distributions, but it is beyond the scope of this work to delve that deep. For those who may be interested more details can be found in Wright and Alison (1984).

What other effect does the age distribution have on kinetic parameters? It should be apparent from Fig. 4.2 that in the steady state the rate of entry (r) into S should equal the rate of entry into M, i.e. $r_S = r_M$, but in the exponential phase there are more cells at the beginning of S than at the beginning of M therefore $r_S > r_M$. In the steady state

$$T_c = \frac{T_S}{\text{LI}} = \frac{T_M}{\text{MI}}.$$

At the other extreme, for a complete exponential age distribution with no cell loss, the cell cycle time is approximated by the relationship

$$T_c = \ln 2\left(\frac{T_S}{\text{LI}}\right) = \ln 2\left(\frac{T_M}{\text{MI}}\right).$$

So use of even the simplest cell kinetic equations can be hazardous without a knowledge of the assumptions behind them. Having examined these, we can now look at some methods for assessing proliferation.

State Measurements

Thymidine Incorporation

The study of cell kinetics really only developed with the ability to specifically label newly synthesised DNA (see Steel 1986). Our knowledge of cell growth control has been greatly enabled by this breakthrough. However, a number of technical points must temper enthusiasm for the method.

Thymidine Metabolism

Most normal cells have the ability to take up exogenously supplied thymidine and incorporate it into their DNA. This is often called the *salvage pathway* as it utilises *free* thymidine rather than relying on thymidine synthesis. Why such a pathway exists is not clear as in many situations free exogenous thymidine is a rarity. The pathway is unique in that no other DNA precursors are solely incorporated into DNA. Uridine, for example will label both DNA and RNA. Such precursors can be used successfully, but only where RNA can be eliminated, e.g. in flow cytometric experiments. Any DNA precursors labelled by the de novo path will also label DNA and RNA. The rate-limiting step in the uptake and incorporation of exogenous thymidine is the enzyme thymidine kinase. Obviously any procedure affecting this enzyme's activity would also affect the uptake of any thymidine supplied. Most thymidine labelling experiments performed rely on the assumption that every DNA-synthesising cell exposed to the precursor will incorporate it in a dose-dependent manner. This is not necessarily the case, for a number of reasons, such as those given below.

Pool Size

The amount of free thymidine in the cytosol is low, but the application of exogenous thymidine will change this. Radioactive thymidine as purchased is a mix of labelled and un-labelled thymidine. The greater the proportion of label the higher the specific activity of the solution is said to be. If labelling is performed with low specific activity, the cell is exposed to abnormally high levels of unlabelled thymidine (often termed *cold thymidine*). This in itself may affect proliferation (Beck 1982; Post and Hoffman 1968). There is also some evidence that some cells may have the ability to retain exogenously supplied thymidine in a form which is not freely exchangeable and to then incorporate the label at a later date (Hume and Potten 1982).

Radiation Damage

If radioactive thymidine is used as a label and the population is studied immediately after application the problems associated with self-irradiation will be limited. However, if high specific activity thymidine is used and/or the population is studied over prolonged periods, nuclear damage may occur.

The most popular label is tritium: its β emissions have a very short pathlength in biological materials, and thus most of the energy will be deposited within the nucleus. This can actually be put to useful effect by haematologists using the thymidine-suicide technique to estimate the number of rapidly proliferating cells in a population. Using high doses, the most active cells become intensely labelled and the self-irradiation causes enough damage for them to die. In most cases, however, self-irradiation is an undesirable effect.

Re-utilization

In vivo the availability of label is short – the unincorporated material is either transported away or broken down. In long-term experiments thymidine may be released from dying or differentiating cells (e.g. red blood cells and squames, both of which are enucleate). This thymidine can be taken up by dividing cells and may give misleading results.

Synthesis Rate Changes

The amount of label taken up by a population of cells within a given time is governed by two factors, the number of cells in DNA synthesis and the rate at which DNA is being synthesised. There is evidence that the rate of synthesis varies through the S phase. Clausen et al. (1980, 1983) have found that in rodent epidermis cells in mid-S are not labelled even by high-specific activity labels. The biological basis for this is unclear.

Other Artifacts

There is the possibility that thymidine may be incorporated into mitochondrial DNA or that DNA repair may be detected rather than scheduled DNA synthesis.

All of these problems, with the exception of radiation damage, also apply to the use of thymidine analogues such as BUdR, which utilise the same pathway.

Non-autoradiographic Method of Detecting Incorporated Thymidine

This is a simple method that has been applied in a variety of tissues. The tissue to be examined is exposed in some way to radiolabelled thymidine, the entire tissue is extracted and the incorporated label measured by scintillation counting. The alternative is to purify DNA from the tissue first; this approach is most often used in in vitro experiments. Unfortunately, there is little doubt that this method has little to recommend it. Insufficient attention has been paid to the numerous potential defects inherent in the technique, which have been repeatedly pointed out (see above, and Maurer 1981) but have been consistently ignored by many workers. It is not the purpose of this chapter to reiterate these reservations at length, except to advise that the method be abandoned by all those with even pretensions towards critical work; the only positive factor in its favour is ease of performance, and it is probably this that has led to its perpetuation and popularity. We would doubt the significance of any investigation where the assay of proliferation rests solely on this method.

Mitotic and Labelling Indices

The simplest method and probably the first ever used was to compare the number of mitoses (see Chap. 5). These can be easily observed even by the inexperienced microscopist. A simple count can be made and different samples compared. In the 1950s, with the advent of radioisotopic labelling, it became possible to label DNA as it was being synthesised (Howard and Pelc 1953; see Steel 1986). The cells that have incorporated the label can be detected by applying a photographic emulsion over tissue sections. When developed, this will reveal grains overlying the nuclei of cells that have incorporated thymidine during the labelling period. These cells can be counted and samples compared. The usual isotope used is tritium (^{3}H): because of its short path-length emissions the grains produced are localised only over DNA, which gives a good image on microscopic examination. Carbon-14 has also been used, but it produces long trails of grains which can even overlie adjacent, non-proliferative cells.

Both these methods raise a number of points which are pertinent to other techniques also. How does one enumerate the results? We can count the number of mitotic or labelled cells, but how do we express that number as a fraction of the cell population? In some tissues this may be extremely difficult if there are multiple cell types present, or if dead or differentiated cells are present. In order to produce a mitotic (MI or I_m) or labelling (LI or I_s) index some workers have not even attempted to consider the problem and have counted labelled cells per microscope field. Obviously changes in the volume, or the cellularity of the tissue, or changes in the ratio of cell types within it could produce different results without underlying change in the proliferative rate.

No single recommendation can be made here as each tissue will require individual consideration, but some attempt must be made to define the population being studied so that meaningful data can be extracted and so that the work of different groups can be compared.

A further practical problem specific to methods of labelling DNA is to ensure that all cells have equal access to and can incorporate the label. In the case of blocks of biopsy tissue incubated with a label for short periods in vitro, it is often the case that intense label is found at the edge of the biopsy but none in the central regions to which the label was unable to diffuse. There are also a number of in vivo examples, in which the labelling pattern is not as expected and some other factors must be considered (Clausen et al. 1980, 1983; Haaskjold et al. 1988; Hume and Potten 1982).

Elsewhere in this volume immunocytochemical and cytochemical methods for assessing proliferation are discussed. There are a number of other methodologies that can yield useful data for some studies but are not suitable for basic assessments of proliferative changes, so these will only be mentioned briefly.

The Fraction Labelled Mitoses Method

The fraction labelled mitoses method (Quastler and Sherman 1959) can give information about the cell cycle time in certain circumstances. However, this procedure has a time scale that makes it inappropriate to monitor short-term changes in proliferative rate, and it is difficult to apply to clinical situations. The confidence limits of the data generated are difficult to calculate, and even where available are of doubtful value in statistical tests. Nevertheless, it has been applied extensively by Steel and co-workers particularly in solid tumours (e.g. Steel 1972, 1977).

The method requires the labelling of a cohort or "window" of cells with a pulse of tritiated thymidine and then following these cells as they proceed around the cycle. Multiple samples are taken with time and, after tissue processing, the fraction of mitoses that is labelled is counted at each point (hence the term FLM, or sometimes PLM for percentage labelled mitoses).

At time zero the only cells with label will be those in S phase, so the FLM will be zero (see Fig. 4.3). After a period equal to that required for a cell that was at the end of S phase when the label was applied to move into mitosis, the curve will begin to rise. This time will give an estimate for the minimum G_2 duration. The flow of labelled cells from S phase continues and thus the FLM will continue to rise and should stay at one for a time equal to the duration of S phase. This is because the labelled window of cells is "S phase wide". The FLM will then drop to zero and will only rise again when the first labelled cell re-enters mitosis in its second cycle. As the curve rises again, the comparison of any two points on the successive curves should give the time to complete an entire cell cycle, the value T_c. The duration of M can be estimated as this should be the time taken for the curve to rise from zero to one (or indeed to fall from one to zero).

Thus the full set of cell cycle parameters can be measured:

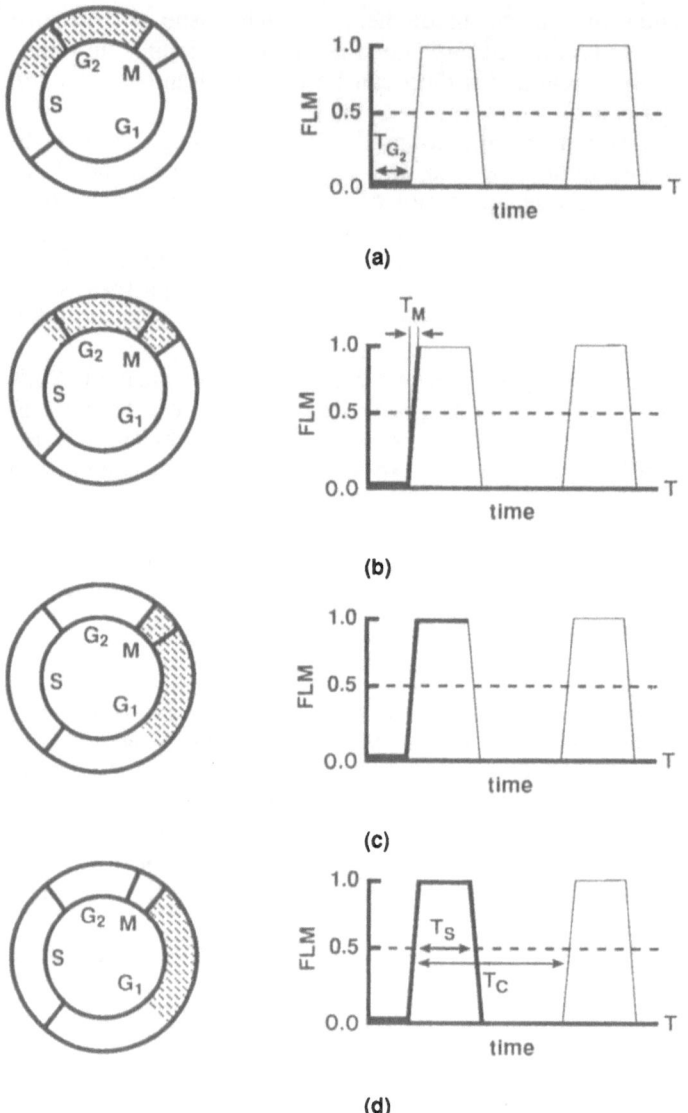

Figure 4.3. An example of a fraction labelled mitosis (FLM) curve. A window of cells is labelled with a short pulse, this window will be equal to an S phase duration in "width". In samples taken at various times after labelling the percentage or fraction of labelled mitotic cells is counted. At the first point, immediately after labelling, no labelled mitoses will be observed. Cells will leave S phase, and at a time equal to the minimum G_2 they will enter mitosis. The FLM curve will then start to rise. As the cohort of labelled cells moves through the cycle a point is reached where all mitotic cells are labelled (**b**). The time between the first rise in the curve and the achievement of an FLM value of one is the duration of mitosis T_M. As the window of labelled cells is S phase "wide" the curve will stay at one for that duration allowing measurement T_S (**c**). As the last labelled cell leaves mitosis the FLM value drops to zero (**d**). The labelled cells pass through G_1, S, G_2 and back into M. At this point the curve begins to rise again. The time between any two comparable points (i.e. events in the cell cycle) is the cell cycle time T_c. As we know T_S, T_{G_2} and T_M, we can calculate the mean G_1 by subtraction from the measured T_c. In practice, most measurements are made at the 50% level. Real FLM curves never match these ideals, and a second peak can often be absent. Mathematical approaches can be used to extract the required information.

The cell cycle time, T_c

The S phase duration, T_S

The G_2 duration, T_{G_2}

The duration of mitosis, T_M

And by subtraction of T_S, T_{G_2} and T_M from T_c we get the duration of G_1, T_{G_1}.

This is fine for an idealised population but in the real world the situation is never as simple and there are a number of problems. The practical problems include the fact that the curves are never as good as the theoretical and in some cases bear no resemblance to it at all! The reasons for this are numerous and include:

1. The accessibility of the labelling agent and its possible toxicity. It is possible that the use of tritiated thymidine could in itself perturb the cell cycle, (Beck 1982; Post and Hoffman 1968)
2. The time scale involved is both a practical problem to apply and makes it unsuitable for the study of rapid changes
3. The method assumes that the cells are asynchronous; this may not be the case and synchrony would have dramatic effects on the FLM curve
4. Damping – the curve will not have sharp peaks as no two sister sibling cells will remain synchronous to the next division; there is a stochastic element, and this will lead to a spread of values, which will damp the curve
5. Re-utilization of the label may occur at later times, particularly in rapid-turnover tissues or where there is a lot of cell death
6. The kinetic organisation of the tissue may also affect the results. In tissues where dividing cells are spatially segregated from the mature cells, the decision as to which cells to count again becomes a problem.

The FLM method is good for assessing the duration of the cell cycle and its phases, but not for measuring the proliferative rate and its changes, particularly where the response being studied may be short lived.

The Continuous Labelling Method

In the continuous labelling method, a label, usually tritiated thymidine, is made available to the cell population over a period of time usually greater than one cell cycle time; this method is considered to give information about the growth fraction, although this is debatable. Ethical considerations alone would often exclude this method from human application.

The method relies on the continuous availability of the label to all the cells in a population which is sampled with time. A curve of the labelling index (LI) is constructed (see Fig. 4.4 for a theoretical example). The curve will start above zero at the flash/pulse labelling level and will rise at a rate equal to the rate of entry into S (see below). At a time equal to the duration of G_2 the curve should double its rate of increase as labelled cells physically divide at mitosis. In practice, the spread of G_2 durations normally masks this, and the curves produced in vivo never show this inflection. In tissues with no cell loss or migration and a growth fraction of one the curve rises to reach a plateau

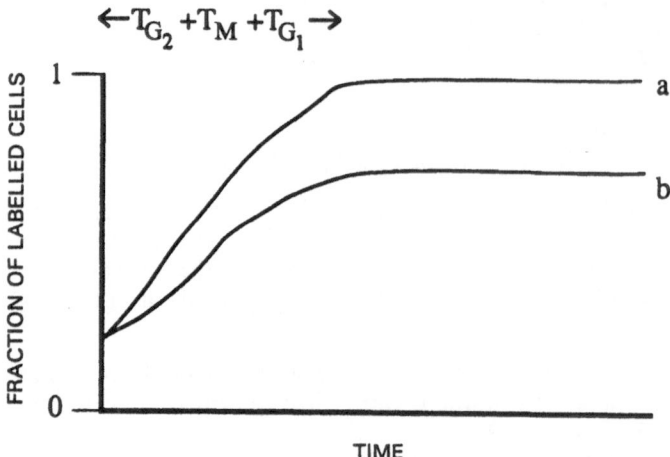

Figure 4.4. A stylised continuous labelling curve: **a** for a population with a growth fraction (I_p) of 1, and **b** for a population with I_p less than 1. The point at which the plateau is reached represents $T_{G_2} + T_M + T_{G_1}$.

at 100%. The time taken to reach this should be equal to the time taken for a cell, which had just exited S phase when the label was applied, to move through G_2, M, and G_1 (and an insignificant fraction of S, sufficient to make it detectable as labelled). In real populations such as renewing tissues and tumours, cell loss and migration do occur and this will affect the shape of the curve, distorting the results. It can be difficult to decide exactly when the plateau is achieved. The level of the plateau should be a measure of the growth fraction, but there are problems: in a renewing situation the plateau consists of cells which are being labelled over and over again. The method is also sensitive to the age distribution of the cells lost from the proliferative compartment. If cells are lost from the non-growth fraction compartment (e.g. post-mitotic transit cells in epithelia) they would not be detected at all by this method. Thus the continuous labelling method can only give reasonably reliable estimates of growth fraction from systems where there is no cell loss and there is only a low level of proliferation, i.e. not for most tumours or most renewing systems.

Thymidine Labelling – Clinical Applications

Since thymidine-based methods require labelled nucleotide to be incorporated into DNA during S phase, they require either the in vivo administration of tritiated thymidine or incubation of a biopsy with thymidine in vitro for some time. Both of these procedures militate against the use of tritiated thymidine in clinical practice, since the former can rarely be justified and the latter is associated with a number of theoretical and practical problems (see above). Nevertheless for many years this was the only available method for assessing cell proliferation, and the data reported in many studies has been pivotal in the development of our understanding of the growth of tumours (Malaise et al. 1973; Steel 1977). Furthermore, in some centres thymidine labelling has been extensively used on clinical material in a wide range of conditions (Malaise et al. 1973; Steel 1977; Wright and Alison 1984), particularly lymphomas (Bremer 1978; Cooper et al. 1976; Peckham and Cooper 1969, 1970) and breast tumours (Meyer et al. 1986).

Studies of breast carcinoma have shown that thymidine labelling correlates closely with a range of pathological and clinical variables, and that it may be a predictor of biological behaviour (Gentili et al. 1981; Meyer et al. 1983, 1986). In other tumours the results of similar studies have been less clear cut (Kerr and Lamb 1988). In a recent study it has been shown that the thymidine labelling index provides prognostic information independent of any other pathological or clinical parameter (Meyer and Province 1988).

Several authors have reported that autoradiographic analysis of thymidine-labelled non-Hodgkin's lymphoma provides information of prognostic value (Brandt et al. 1981; Costa et al. 1981; Silvestrini et al. 1977), and that despite its technical shortcomings measurement of tritiated thymidine by scintillation counting seems to provide similar information (Kvaloy et al. 1981, 1985). A particular advantage of thymidine-based methods is the possibility of deriving rate data. For example in a study of Burkitt's lymphoma, Iversen et al. (1974) were able to show that in tumours with a mean clinical doubling time of 66 h, with a labelling index of 17%, 2.7% of cells entered DNA synthesis per hour and the cell cycle time was 25.7 h. From these experiments it was shown that more than two-thirds of cells were lost (cell loss factor 69%). These observations fit with the clinical picture of a rapidly growing tumour, the histological features (numerous mitotic figures and apoptotic bodies) and the response to aggressive chemotherapy. Similar data have accrued in a number of other studies of lymphoma and other neoplasms.

Flow Cytometry

The use of a flow cytometer (see Chaps. 6 and 7) is one answer to the problem of the time-consuming microscopy required by labelling methods described so far. If the tissue can be reduced to a single-cell suspension this can be a very useful method; for this reason much of the work has been performed on blood cells. The sample is reacted with a dye that binds stoichiometrically to DNA. The single-cell suspension is passed through a laser beam and the dye emits a signal that can be measured. A large number of nuclei, typically 10 000 or more, can be analysed in minutes. The data are usually expressed as a frequency curve of DNA content. This gives an estimate of DNA per cell. Various mathematical transforms can be applied which will yield estimates of the number of cells with 2C, intermediate or 4C DNA content equating to the G_0/G_1, S and G_2+M compartments. Therefore estimates of the number of cells in S phase can be obtained and samples may be compared. This sounds too good to be true – large sample size, automated analysis and rapid results! Technical difficulties apart, the machines remain expensive, although bench-top analysers are now available at a more reasonable price. The problem remains of what the information obtained means: the naïve presumption is that a larger percentage of cells in S phase indicates that the population is dividing faster. Not so; it means that more cells are in S phase. It is theoretically possible for one population, to have a cycle time of say 50 h, and another population to have a cycle time of 25h; if the duration of each phase of the cycle was in exact proportion, the flow cytometer would not detect any difference between the populations as the distribution around the cycle would be exactly the same. This is the fundamental problem of *state measurements*: they only indicate what the population is doing at any one time, not the flux of cells. The cytometer, when combined with other techniques, can be used to measure fluxes (see Chap. 7), but these methods have been used relatively infrequently and simple DNA histograms are the norm.

Summary

The advantages of these methods are that they are usually simple to apply and to analyse. The disadvantages are numerous. On the practical side, the use of radioisotopes that can be incorporated into DNA is a problem for in vivo studies. The main failing is that they produce data which can be equivocal. There is no doubt that in many cases two populations with different labelling indexes will be proliferating at different rates, but these methods do not allow the investigator to be *sure* that they are different, and it is bad science to make the assumption. However, problems apart, the mitotic and labelling indices can be useful guides to the proliferative activity of cell populations. The applications of immunohistological methods are rapidly becoming widespread tools for the characterisation of proliferation in clinical material (see Chap. 8) but suffer from many of the theoretical and practical limitations described above.

Rate Measurements

There has been some confusion in the terminology of cell proliferation: the term "proliferative rate" is a generic one which usually encompasses all factors which relate to the speed of production of new cells. Perhaps the most important parameter is the *birth rate* or *cell proliferation rate* which is the rate at which new cells are being made. If we could measure the cell production rate, in both control and experimental situations, then we would have gone a long way towards solving our problem. However, very few of the measurements available relate even tenuously to the cell production rate.

The difference between the information gleaned from state and rate measurements can be highlighted by a simple analogy. A person interested in road traffic might make a state measurement by looking at a fixed length of road and counting the number of vehicles on it at a given instant at several particular times during the day. He might observe that at some instant during the morning rush hour there are many more vehicles on the road than at some instant during the night. The assumption from a state point of view is that the traffic in the morning moves faster because there are more vehicles per unit length of road measured. The rate point of view would differ, in that a measurement of the number of vehicles passing a particular point on the road *during some defined period of time* would reveal that in the morning the vehicles are passing only slowly as they are in a traffic jam. At night they pass quickly as they are unimpeded. Clearly both sets of data contain potentially useful information, but state data is quite different from rate data! If we are asking questions about *how fast* we need to make measurements of *rate* not *state*.

At a fundamental level, the two parameters that jointly determine the cell production rate are the *cell cycle time*, the interval between successive cell divisions, and the *growth fraction*, which is the proportion of cells which contribute to new cell production. Unfortunately, measurement of both of these parameters is a difficult, expensive and frequently unrewarding exercise, and employs techniques that are really unsuitable for the question we are asking, i.e. how do we detect differences in proliferative rates?

Two rate measurements are generally available: *the rate of entry or flux into DNA synthesis* and the *rate of entry or flux into mitosis*.

The Rate of Entry into S Phase

The rate of entry into the S phase is usually measured by an appropriate double-labelling method. The rationale is to label a cohort of epidermal cells in S first with an application of labelled DNA precursor and, after an appropriate time (usually 1 h, but see later in this section), a further application labels those cells that have entered S in the time elapsed since the first application; the difference between the two labelling indices gives the flux into S during the time between the two injections. The main problem therefore is to isolate those cells labelled by the first application from those labelled by the second: there are three main ways of doing this:

1. By far the best method is to use thymidine labelled with isotopes of differing energies. The usual procedure is to use ^{14}C-labelled and tritium ^{3}H-labelled thymidine: ^{14}C has a higher energy than ^{3}H, and can be detected by its higher penetration power; the section is therefore coated with a thick layer of emulsion, or preferably with two layers of emulsion (Schultze et al. 1976). The lower energy ^{3}H particles are detected by silver grains in the lower emulsion, while the higher energy ^{14}C reaches the uppermost emulsion layer. This procedure allows the separation of ^{3}H- and ^{14}C-labelled cells, and also those labelled with both ^{3}H and ^{14}C, and both influxes and effluxes from the S phase can be calculated (Burholt et al. 1976). It is important to exert careful control over the specific activities and doses of the isotopes (Schultze et al. 1976), and over the autoradiographic procedure. An alternative approach is to use tritiated thymidine and BUdR.

2. A second method of separating the labelled cohorts is to use two applications of tritiated thymidine of differing doses, or concentration, which should result in a lightly labelled and a heavily labelled cohort, theoretically distinguishable by the grain count. It is important to ensure that the distributions of grain counts over the respective cohorts do not overlap.

3. A third procedure is to give two applications of the same isotope, separated in time, and merely to measure the two labelling indices to obtain the difference. One problem here is that of course only a small difference in

labelling index is being sought, which could lead to problems in interpretation (Ralfs et al. 1981). However, this difference can be augmented by increasing the interval between the applications: this can only be done if cells that pass through G_2 and M are prevented from dividing and artificially elevating the first labelling index. This can be achieved by giving a metaphase-arrest agent, such as demecolcine or colchicine, at the time of the first injection (Tvermyr 1972).

This measurement, the flux into DNA synthesis, is valuable in that it should reflect the cell production rate. However, to equate the flux into S with the cell production rate we need to assume that there is no effective cell loss between the beginning of the S phase and cell division itself, and this may not be so in some cases (Ralfs et al. 1981). The flux into S is an important technique in the context of the measurement of proliferative rate. Nevertheless, for clinical work the use of isotopes is a problem, as is the accessibility of the labelling agent.

The Rate of Entry into Mitosis

From a theoretical point of view, the rate of entry into mitosis should be the ideal measurement, since for every cell which undergoes mitosis there will be a net gain of one cell if there is no change in cell loss rates. To measure the rate of entry into mitosis is to measure the rate at which new cells are "born", and this is often called the *birth rate*. This method (termed metaphase arrest or stathmokinesis) uses the ability of certain compounds to arrest cells in metaphase by disrupting or preventing the formation of the mitotic spindle. Practically speaking, the agent is administered systemically, and serial measurements are made of the mitotic or metaphase index with time after injection. The resulting linear rate of metaphase accumulation is usually found by a least-squares fit, giving the cell-production rate with its confidence interval. The general advantages and drawbacks of the method have been discussed in detail by Wright and Appleton (1980), but important prerequisites include: (a) the use of optimal dosage; (b) the proof of linearity of metaphase collection; (c) the avoidance of a delay period before metaphase arrest becomes complete, and also of metaphase degeneration; and (d) the choice of an appropriate line-fitting procedure.

This method is valuable, and may be the method of choice in accessible areas such as the epidermis of experimental animals, but in normal human epidermis, for some reason, there is no effective rise in the metaphase index over the experimental period, which is usually 3–4 h (Camplejohn et al. 1981); why this should be is obscure. It could be because of the very slow proliferative rate in normal human epidermis and the fact that the consequent small increase over the experimental period is inadequate for the detection limits of the method.

A recent development has made rate parameter measurements possible by flow cytometry. This involves using a marker to label a "window" of cells and follow their progression relative to the other G_1- , S- and G_2+M-phase cells. Initially tritiated thymidine was used (Gray et al. 1977), but the technique was difficult to apply; now BUdR is used (see Chap. 7).

Summary

We can now see that rate measurements are the best way to answer the question we first posed. It is clear that they are often impossible to apply in vivo in the clinical situation, although some tissues may be amenable to such measurements, e.g. the epidermis and appendages.

Conclusion

The researcher wishing to study proliferation in clinical material should by now be in a quandary. The ideal measurement is virtually impossible in the clinical setting and the methods that can be applied may not give the right answer; what can be done? Hopefully the rest of this book may provide some clues! The single most important point is to understand exactly what the method applied actually tells you, and with what confidence limits. It is also essential that the limitations of the methodology are understood and that the assumptions made in reaching the conclusion are clearly stated.

It might seem that some of the points raised are academic and of no relevance to most people. It is clear that only by a careful study of proliferation have we gained insight into the proliferative organization of tissues (see Chap. 3), and if any sort of therapy is to be based on the results obtained or if the data are to be used as a prognostic indicator then it is important to be sure that the data are correct. Similarly, many simple methods may not detect differences which only more sophisticated studies would reveal. There are a number of new methods which have varying merits but all correlate in some way with proliferative activity. In time, some will fall by the wayside and some may prove very useful. The remaining chapters should help to steer the reader through the potential minefield of proliferative measurements.

References

Aherne WA, Camplejohn RS and Wright NA (1977) An Introduction to cell population kinetics. Edward Arnold, London

Beck H-P (1982) Radiotoxicity of incorporated ^{3}H-thymidine consequences for the interpretation of FLM data of irradiated populations. Cell Tiss Kinet 15:469–480

Brandt L, Olsson H and Monti M (1981) Uptake of thymidine in lymphoma cells obtained through fine-needle aspiration biopsy. Relation to prognosis in non-Hodgkin's lymphoma. Eur J Cancer 17:1229–1233

Bremer K (1978) Cellular renewal kinetics of malignant non-Hodgkin's lymphomas. In: Mathe G, Seligmann M and Tubiana M (eds) Lymphoid neoplasias II. Clinical and therapeutic aspects. Springer Verlag, Berlin, pp 5–11

Burholt DR, Schultze B and Maurer W (1976) Mode of growth of the jejunal cryp cells of the rat; an autoradiographic study using double labelling with ^{3}H and ^{14}C thymidine in lower and upper parts of the crypts. Cell Tissue Kinet 9:107–111

Camplejohn RS, Gelfant S and Chalker D (1981) An attempt to use vincristine and colcemid to measure proliferative rates in normal human epidermis in vivo. Br J Dermatol 104:243–248

Clausen OPF, Thorud E and Bolund L (1980) DNA synthesis in mouse epidermis: labelled and unlabelled basal cells in S phase after administration of tritiated thymidine. Virchows Arch B Cell Pathol 34:1–11

Clausen OPF, Elgjo K, Kirkhus B, Pedersen S and Bolund L (1983) DNA synthesis in mouse epidermis: S phase cells that remain unlabelled after pulse labelling with DNA precursors progress slowly through S. J Invest Dermatol 81:545–549

Cooper EH, Peckham MJ, Millard RE, Hamlin IME and Gerard-Marchant R (1976) Cell proliferation in human malignant lymphomas. Analysis of labelling index and DNA content in cell populations obtained by biopsy. Eur J Cancer 4:287–296

Costa A, Bonadonna G, Villa E, Valagussa P and Silvestrini R (1981) Labelling index as a prognostic marker in non-Hodgkin's lymphoma. J Natl Cancer Inst 66:1–5

Gentili C, Sanfilipo O and Silvestrini R (1981) Cell proliferation and its relationship to clinical features and relapse in breast cancers. Cancer 48:974–979

Gray JW, Carver JH, George YS and Mendelsohn MC (1977) Rapid cell cycle analysis by measurement of the radioactivity per cell in a narrow window in S phase (RCS). Cell Tissue Kinet 10:97–109

Haaskjold E, Refsum SB, Bjerknes R and Paulsen TO (1988) The labelling index is not always reliable. Discrepancies between the labelling index and the mitotic rate in the rat corneal epithelium after intraperitoneal and topical administration of tritiated thymidine and colcemid. Cell Tissue Kinet 21:389–394

Howard A and Pelc SR (1953) Synthesis of deoxyribonucleic acid in normal and irradiated cells and its relationship to chromosome breakage. Heredity, Suppl 6:261–273

Hume WJ and Potten CS (1982) A long lived thymidine pool in epithelial stem cells. Cell Tissue Kinet 15:49–58

Iversen OH, Iversen U, Ziegler JL and Bluming AZ (1974) Cell kinetics in Burkitt's lymphoma. Eur J Cancer 10:155–163

Kerr KM and Lamb D (1988) A comparison of patient survival and tumour growth kinetics in human bronchogenic carcinoma. Br J Cancer 58:419–422

Kvaloy S, Godal T, Marton PF, Steen H, Brennhovd IO and Abrahamsen AF (1981) Spontaneous ^{3}H thymidine uptake by histological subgroups of human B cell lymphomas. Scand J Haematol 26:221–234

Kvaloy S, Kaalhus O, Hoie J, Abrahamsen A and Godal T (1985) ^{3}H thymidine uptake in B cell lymphomas. Relationship to treatment response and survival. Scand J ..aematol 34:429–435

Lajtha LG (1963) On the concept of the cell cycle. J Cell Comp Physiol 60 (Suppl 1):143–145

Malaise EP, Chavaudra N and Tubiana M (1973) The relationship between growth rate, labelling index and histological type of human solid tumours. Eur J Cancer 9:305–312

Maurer HR (1981) Potential pitfalls of ^{3}H-thymidine techniques to measure cell proliferation. Cell Tissue Kinet 14:111–120

Meyer JS and Province M (1988) Proliferative index of breast carcinoma by thymidine labelling: prognostic power independent of stage, estrogen and progesterone receptors. Breast Cancer Res Treat 12:191–204

Meyer JS, Friedman E, McCrate MM and Bauer WC (1983) Prediction of early course in breast carcinoma by thymidine labelling. Cancer 51:1879–1886

Meyer JS, Prey MU, Babcock DS and McDivitt RW (1986) Breast carcinoma cell kinetics, morphology, stage and host characteristics. A thymidine labelling study. Lab Invest 54:41–51

Peckham MJ and Cooper EH (1969) Proliferation characteristics of the various classes of Hodgkin's disease. Cancer 24:135–146

Peckham MJ and Cooper EH (1970) The pattern of cell growth in reticulum cell sarcoma and lymphosarcoma. Eur J Cancer 6:453–463

Post J and Hoffman J (1968) Early and late effects of ^{3}H-TdR-labelled DNA on ileal cell replication in vivo. Radiat Res 34:570–581

Quastler H and Sherman FG (1959) Cell population kinetics in the intestinal epithelium of the mouse. Exp Cell Res 17:420–438

Ralfs I, Dawber RPR, Ryan TJ, Dufhill M and Wright NA (1981) The kinetics of metaphase arrest in human psoriatic epidermis; an examination of optimal experimental conditions for determining the birth rate. Br J Dermatol 104:231–242

Schultze B, Maurer W and Hagenbush H (1976) A two emulsion autoradiographic technique and the discrimination of the three different types of labelling after double labelling with ^{3}H and [14]C Thymidine. Cell Tissue Kinet 9:245–255

Silvestrini R, Piazza R, Riccardi A and Rilke F (1977) Correlation of kinetic findings with morphology of non-Hodgkin's lymphomas. J Natl Cancer Inst 58:499–504

Steel GG (1968) Cell loss from experimental tumours. Cell Tissue Kinet 1:193–207
Steel GG (1972) The cell cycle in tumours: an examination of data gained by the technique of labelled mitoses. Cell Tissue Kinet 5:87–100
Steel GG (1977) The growth kinetics of tumours. Oxford, Clarendon Press
Steel GG (1986) Autoradiographic analysis of the cell cycle: Howard and Pelc to the present day. Int J Radiat Biol 49:227–235
Tvermyr EMF (1972) Circadian rhythms in hairless mouse epidermal DNA synthesis as measured by double labelling with ^{3}H-Thymidine. Virch Arch B Cell Pathol 11:43–56
Wright NA and Alison MR (1984) The biology of epithelial cell populations. Clarendon Press, Oxford
Wright NA and Appleton DR (1980) The metaphase arrest method – a critical review. Cell Tissue Kinet 13:643–658

5 Mitosis Counting

C.M. QUINN and N.A. WRIGHT

Introduction

It is a truism to say that the main (nay, only) method of assessing cell proliferation which has enjoyed any routine popularity has been the assessment of mitotic activity. This is obviously because nothing special has to be done to the tissues (indeed routine sections are usually used). However, instead of approaching such measurements on an appropriate basis, i.e. achieving measurements of accuracy and known precision, most workers have relied on easy, non-critical, rule-of-thumb assessments. Such an approach is, in our opinion, a perfect recipe for disaster. In this chapter we review the several methods of assessing mitotic activity, the purported role of these measurements in prognosis, and reflect on the vagaries of these measurements.

Mitosis Counting – Methods

Mitotic Count

The mitotic count is defined as the number of mitoses per 10 HPF, derived by simple observation. This method is rapid and easy – ideal for incorporation into routine diagnostic histopathology. It is not, however, a standardised method (Quinn and Wright 1990; Sadler and Coghill 1989): it takes no account of cell size (Akerman et al. 1987; Donhuisjen 1986; Wright 1984) and the area of a single HPF may vary up to six-fold between different microscopes (Ellis and Whitehead 1981). The mitotic count is also subject to inter-observer variation (Silverberg 1976). Despite these obvious limitations the mitotic count is widely used in clinical practice. Most studies reporting the value of mitosis counting in differentiating benign and malignant conditions have used the mitotic count rather than the alternative methods of mitosis index estimation. In this regard, it is well to be aware that terminology is sometimes confused in the literature, with the term mitotic index being used instead of mitotic count.

Mitotic Index

The mitotic index is the number of mitoses expressed as a percentage of the number of interphase nuclei counted, usually 1000. The mitotic index overcomes the problems of variation in cell size and HPF area. It is a standardised method derived by simple observation and is thus more acceptable than the mitotic count. It is of course more time-consuming and is not so popular in routine practice.

Mitotic Rate

The mitotic rate is defined as the rate at which cells are entering mitosis, expressed as a percentage of cells counted, per hour (Wright 1984). The mitotic rate does not depend on the duration of mitosis and is therefore a more independent variable than the count or index. The mitotic count and mitotic index depend not only on the number of cells in mitosis but also on the duration of mitosis in the cell population being studied. Large numbers of mitoses may be due either to rapid proliferation or to a long mitotic phase.

The mitotic rate can be measured in human tissue by arresting cells during mitosis and measuring their accumulation. This is achieved by administering compounds such as colchicine, colcemid or vinca alkaloids, either systemically (Camplejohn et al. 1973; Wright et al. 1979) or by local injection (Duffill et al. 1976; Ralfs et al. 1981). These compounds arrest cells in metaphase and the technique is thus known as the metaphase arrest technique (Wright and Alison 1984; Wright and Appleton 1980). Following administration of the compound the tissue of interest is biopsied at intervals. The number of arrested metaphases is counted in paraffin tissue sections and the incidence plotted against time. A straight line is obtained, the slope of which is equal to the rate of entry of cells into mitosis – the mitotic rate (also known as the birth rate). This method has been used in clinical situations, notably for measuring the birth rate in human tumours (Camplejohn et al. 1973; Taylor et al. 1977; Wright et al. 1977). It is an accurate parameter of cell proliferation but is a lengthy procedure requiring ethical consent.

Mitosis Counting – Applications

Uterine Smooth Muscle Tumours

Tumours of smooth muscle constitute the most common neoplasms of the uterus. A large percentage of these tumours are benign with the ratio of benign : malignant tumours estimated at 800 : 1 (Zaloudek and Norris 1987). Benign and malignant smooth muscle tumours of the uterus differ both clinically and pathologically. Leiomyomas classically occur during the reproductive years while leiomyosarcomas tend to occur in postmenopausal women. Leiomyomas are usually multiple, well-circumscribed and cytologically innocuous in contrast to malignant lesions, which are frequently solitary, ill-defined and necrotic with marked cytological atypia. These distinctions are not invariable, and the distinction between a benign and malignant smooth muscle tumour may prove histologically difficult. It is in the differentiation of these two conditions that the mitotic count is most commonly used in diagnostic histopathology. Despite its popularity, the validity of the mitotic count as a diagnostic and prognostic parameter in uterine smooth muscle tumours is not universally accepted.

Many authorities advocate the mitotic count as the most reliable method of distinguishing benign and malignant uterine smooth muscle tumours (Christopherson et al. 1972; Gallup and Corday 1979; Hart and Billman 1978; Kempson 1976; Kempson and Bari 1970; Marchesse et al. 1984; Norris 1976; Taylor and Norris 1966), but there has been some disagreement regarding the number of mitoses per 10 HPF required to warrant a diagnosis of malignancy. It has recently been proposed that tumours with a mitotic count of 10 or more mitoses per 10 HPF, and tumours with a count between 5 and 9 mitoses per 10 HPF accompanied by moderate to severe cytological atypia, are to be regarded as malignant. Tumours with a count between 5 and 9 mitoses per 10 HPF with mild cytological atypia are denoted "smooth muscle tumours of uncertain malignant potential" (Zaloudek and Norris 1987). Cellular and bizarre leiomyomas have a mitotic count of less than 5 mitoses per 10 HPF (Christopherson et al. 1972; Hart and Billman 1978; Kempson and Bari 1970; Marchesse et al. 1984; Zaloudek and Norris 1987). The importance of adequate sampling in assessing tumour mitotic activity is emphasised by all proponents of the mitotic count.

Not all workers agree that the mitotic count is the major prognostic criterion in uterine smooth muscle tumours. In a study of mesenchymal tumours of the uterine corpus, Saksela et al. (1974) reported a good correlation between mitotic activity and tumour behaviour in leiomyosarcoma but found the extent of tumour spread at the time of surgery to be the main prognostic indicator. Hannigan and Gomez (1979) shared this view and found no number of mitoses below which the diagnosis of leiomyosarcoma was excluded. Burns et al. (1979) reported increased mitotic activity with aggressive tumour behaviour but concluded that no single microscopic feature was uniformly predictive of tumour behaviour. Silverberg (1976) expressed considerable concern about the use of the mitotic count as the sole indicator of malignancy in uterine smooth muscle tumours, mainly because of the problem of interobserver variation in mitosis counting. In an earlier study he found the menstrual status of the patient and the macroscopic appearance of the tumour to be the most valuable indicators of prognosis (Silverberg 1971). He and others favour a multifactorial approach based on clinical and pathological features in the diagnosis of uterine smooth muscle tumours (Perrone and Dehner 1988; Silverberg 1971).

The mitotic count appears to be unreliable in predicting the behaviour of epithelioid smooth muscle tumours of the uterus. Epithelioid tumours with a mitotic count in excess of 5 mitoses per 10 HPF are regarded as malignant (Kurman and Norris 1976; Zaloudek and Norris 1987), but a count of less than 5 does not exclude the possibility of malignant behaviour (Bucsema et al. 1986; Kurman and Norris 1976; Zaloudek and Norris 1987). On a similar note, the smooth muscle tumour variant known as myxoid leiomyosarcoma is notoriously aggressive despite the frequent finding of low mitotic activity (King et al. 1982; Zaloudek and Norris 1987).

Gastrointestinal Smooth Muscle (Stromal) Tumours

Smooth muscle tumours of the gastrointestinal tract occur much less frequently than their uterine counterparts. These tumours display variable biological behaviour which, together with their relative infrequency, can make differential diagnosis and prognostication difficult. The mitotic count has been advocated as a useful but not totally reliable criterion in predicting the likely outcome of these tumours (Appelman and Helwig 1976, 1977; Evans 1985; Ranchod and Kempson 1977). In a series of 100 gastrointestinal smooth muscle tumours, all tumours with a mitotic count in excess of 5 mitoses per 10 HPF behaved aggressively, but as many as 40% of biologically malignant tumours had counts lower than 5 mitoses per 10 HPF (Ranchod and Kempson 1977). In a series of 49 cellular leiomyomas, one tumour with a count less than 5 mitoses per *50* HPF pursued a metastatic course (Appelman and Helwig 1977). Thus, the degree of mitotic activity appears to be useful in delineating high-grade leiomyosarcomas but not so in the separation of leiomyomas from low-grade malignant lesions (Evans 1985).

Endometrial Stromal Sarcomas

Endometrial stromal sarcomas (ESS) are classified into high-grade stromal sarcoma (HGSS), characterised by rapid progression and a high mortality rate, and low-grade stromal sarcoma (LGSS), which is locally aggressive and colonises lymphatics but generally pursues a more favourable clinical course (Zaloudek and Norris 1987). The LGSS is also known as endolymphatic stromal myosis and sometimes as endometrial stromatosis.

The distinction between high- and low-grade lesions is made histologically and the degree of mitotic activity is frequently used to differentiate the two conditions. The figure of 10 mitoses per 10 HPF has been proposed as the absolute criterion for the diagnosis of high-grade ESS. Tumours with a mitotic count of 10 mitoses per 10 HPF or greater are designated HGSS, and those with a count less than 10 mitoses per 10 HPF as LGSS (Kempson and Bari 1970; Norris and Taylor 1966; Piver et al. 1984). It has even been suggested, *on the basis of two cases*, that mitosis counting offers a clear distinction between grades of ESS on cytological preparations (Morimoto et al. 1982). Other workers consider the degree of mitotic activity to be the single most useful but not the sole criterion in separating high- and low-grade lesions and recognise cellular atypia as an additional useful feature (Yoonessi and Hart 1977).

The value of the mitotic count in endometrial stromal sarcoma has been challenged. In a series of five cases of endometrial stromatosis (LGSS), two of which had a fatal outcome, Gitstein et al. (1980) found no variation in mitotic activity from case to case. Cytological anaplasia proved to be a more useful indicator of biological behaviour. Similarly, in 33 cases of endometrial stromatosis, Thatcher and Woodruff (1982) found that lack of differentiation of the tumour cell, rather than mitotic activity, was the most significant criterion of malignancy. In the Mayo Clinic review of 24 cases of high-grade ESS, overall survival was significantly related to tumour size, extent of disease and histological grade, but not to the number of mitoses (De Fusco et al. 1989). In a comparative study of low- and high-grade lesions, Evans (1982) found the mitotic count to be of no prognostic significance. He proposed that the diagnosis of low-grade lesions be based on evidence of endometrial stromal differentiation and not on degree of mitotic activity (Evans 1982). Applying this diagnostic criterion to their group of nine patients, Taina et al. (1989) noted that three cases diagnosed as HGSS according to the mitotic count would now belong to the low-grade group. It is of interest that these three patients exhibited prolonged survival times.

These latter studies, together with the shortcomings of the mitotic count (see below), suggest that other histopathological criteria, such as cytological atypia and the absence of stromal differentiation, are more accurate in distinguishing HGSS.

Placental Site Trophoblastic Tumour

This is the condition originally known as "trophoblastic pseudotumour" on the assumption that all cases were benign (Kurman et al. 1976). It has since become known that this condition may pursue a malignant course (Twiggs et al. 1981). In 1981 the condition was renamed placental site trophoblastic tumour (PSTT) to emphasise its malignant potential (Scully and Young 1981). The distinction between benign and malignant variants of PSTT is difficult. At present, the mitotic count is deemed to be the most reliable criterion in predicting the possible outcome of a given tumour. In benign lesions the mitotic count is reported to be less than 2 mitoses per 10 HPF, with most malignant lesions displaying a count in excess of 5 mitoses per 10 HPF (Lathrop et al. 1988; Scully and Young 1981; Young and Scully 1984). However, two cases with low mitotic counts have been reported, one of which demonstrated a fatal outcome (Gloor et al. 1983), the other being associated with extensive pelvic infiltration (Eckstein et al. 1982). Thus, as with other tumours, the validity of the mitotic count as a prognostic indicator is highly questionable. We support the view that hysterectomy remains the treatment of choice in the management of PSTT (Eckstein et al. 1985; Lathrop et al. 1988).

Ovarian Neoplasms

In the assessment of ovarian neoplasms, the problem facing the pathologist is not so much the differentiation of benign and malignant lesions but rather the task of prognosticating on malignant tumours which display notoriously heterogeneous biological behaviour. The conventional parameters used to predict the likely prognosis of these tumours, including clinical stage, tumour grade, histological subtype, and the volume of residual tumour following surgery (Hacker et al. 1983; Malkasian et al. 1984; Richardson et al. 1985; Sorbe et al. 1982) are frequently criticised because of the subjective way in which they are determined (Baak et al. 1982, 1986; Friedlander 1984). In the search for more accurate, objective and reproducible parameters, flow cytometry has emerged as a useful prognostic tool in ovarian cancer. Mitosis counting, using morphometric methods, alone (Haapasalo et al. 1989a, 1990) or in combination with flow cytometric data (Baak et al. 1987, 1988), may also provide helpful prognostic information. One study has shown that mitotic activity taken together with volume density of tumour cells was as good as flow cytometry in identifying favourable and unfavourable patients with advanced ovarian cancer (Rodenburg et al. 1988). The presence of mitotic activity in peritoneal implants from ovarian borderline tumours adversely affects prognosis (Bell et al. 1988).

Lymphomas

Non-Hodgkin's lymphoma is a collective term for a wide range of malignant lymphoid tumours. Like ovarian lesions, these tumours display dramatic variations in biological behaviour, and again the task facing the pathologist is primarily one of prognostication, particularly with regard to choice of therapy. Numerous classification systems, based on histopathological characteristics, have been proposed (Bennett et al. 1974; Gerard-Marchant et al. 1974; The Non-Hodgkin's Lymphoma Pathologic Classification Project 1982; Rappaport 1966; Stansfeld et al. 1988). Assessment of mitotic activity alone appears to be a useful predictor of survival (Donhuisjen 1987). The degree of mitotic activity may offer additional prognostic information to the Kiel classification, particularly in the centroblastic–centrocytic variety (Akerman et al. 1987; Griffin et al. 1988). A high mitotic count in malignant lymphoma, small lymphocytic type delineates a subgroup of patients with a particularly poor prognosis (Evans et al. 1978). This is also the case in centrocytic lymphoma (Swerdlow 1983).

Other Applications

The mitotic count may help to distinguish between ductal carcinoma in situ and the more recently recognised lesion of atypical hyperplasia of the breast (De Potter et al. 1987). Assessment of mitotic activity may be useful in separating cervical lesions caused by human papilloma virus (HPV) types 6 and 11 from lesions caused by HPV types 16, 18 and 33 (Fu et al. 1988). Mitotic activity appears to be a significant predictor of survival in stage-1 cutaneous malignant melanoma (Ronan et al. 1988).

Mitosis Counting – Advantages

The major advantage of mitosis counting is that all parameters (count, index and rate) are derived by simple observation of mitotic figures in paraffin tissue sections and the technique does not require expensive or sophisticated equipment. Estimation of the mitotic count is rapid, hence its popularity in daily diagnostic histopathology. The mitotic index is a more accurate parameter of cell proliferation and, although more time-consuming than the mitotic count, is also suitable for incorporation into routine work. Calculation of the mitotic rate requires patient cooperation and is thus a more lengthy procedure than the mitotic index or count.

Mitosis Counting – Limitations

Tumour Heterogeneity

Mitosis counting is most frequently used in an attempt to predict the behaviour of human tumours. It must be appreciated that tumours do not constitute a homogeneous proliferative population. Wide regional variation in tumour proliferative indices has been reported in animals, depending on the relationship of the sample studied to the periphery of the tumour (Aherne et al. 1977a; Hermens et al. 1969) and to tumour blood vessels (Tannock 1968, 1970) and thus to oxygen concentration (Aherne et al. 1977b). This also applies to human tumours and the mitotic count/index may vary widely between different blocks of the same tumour. Thorough sampling is of obvious import in this regard.

Delay in Fixation

Delay in tissue fixation may alter the number of mitotic figures available for counting in paraffin sections (Donhuijsen et al. 1990; Edwards and Donalson 1964; Graem and Helweg-Larsen 1979; Start et al. 1990). It is suggested that cells in early mitosis may complete and exit the mitotic phase during the fixation process, while few cells enter mitosis following removal of tissue from the body. This problem is especially relevant in the case of large tumour specimens where the fixative may take a long time to reach the centre of the tumour. Prompt fixation, assisted by tumour slicing prior to fixation, will help to halt mitosis in surgical specimens and so preserve mitotic figures for more accurate counting.

Inter-Observer Variation

Inter-observer variation in mitosis counting is a well-recognised problem (Burns et al. 1979; Silverberg 1976). Proposed reasons for such variation include failure to locate the most active region on the slide, failure to distinguish between mitotic figures and pyknotic nuclei, and variation in the number of fields studied (Silverberg 1976). Additional factors include variation in section thickness and in the area studied by different observers. Inter-observer variation may account, in part at least, for the failure of mitotic count proponents to agree on the number of mitoses per 10 HPF necessary to make a definite diagnosis of malignancy (Silverberg 1976). Reproducibility in mitosis counting can be improved by adequate tumour sampling, by applying strict criteria for inclusion as mitotic figures, and by standardising section thickness,

HPF area, etc. (Baak 1990; Baak et al. 1989). Giemsa preparations (Donhuijsen 1986) and certain modified silver stains (Busch and Vasko 1988) may assist in easier recognition of mitotic figures. Results, however, are still greatly affected by the experience of the observer (Donhuijsen 1986), and it would be impossible to eliminate the problem of inter-observer variation altogether.

Cell Size Variation

The mitotic count takes no account of cell size (Wright 1984; Donhuijsen 1986; Akerman et al. 1987). A HPF of a large-cell tumour will contain fewer cells and proportionately fewer mitotic figures than the same HPF of a small-cell tumour. Thus to compare mitotic counts of different tumours may frequently be meaningless. This problem is overcome by using the mitotic index or the mitotic rate as alternative methods of mitosis counting.

Variation in the Area of High-Power Fields

The validity of the mitotic count is seriously compromised by the observation that the area of a single HPF may vary up to six-fold between different microscopes (Ellis and Whitehead 1981). This means that the count for the same tumour assessed by different instruments might vary from 3 to 18 mitoses per 10 HPF, cutting right across the proposed critical level for the diagnosis of leiomyosarcoma. It has been recommended that the count be expressed per square millimetre, in an attempt to standardise results (Hemdal and Frankendal 1985; Tiltman 1988), but this does not eliminate the problem of variation in cell size between different tumours. Alternatively, the area of the HPF can be measured, and the mitotic count standardised to this area, thus allowing comparison between different observers. In this respect, Haapasalo et al. (1989b) have introduced a measurement called the *volume-corrected mitotic index* (M/V index), a measurement designed to give a mitotic count per square millimetre of *neoplastic* tissue. This method has been claimed to be a better prognosticator than the mitotic count in ovarian cancer (Haapasalo et al. 1990) and also claimed to give useful results in the prognosis of pancreatic (Lipponen et al. 1990a) and bladder (Lipponen et al. 1990b) cancer. However, such a volume-corrected mitotic index does not eliminate subjective bias, lacks statistical discrimination, and is by no means new.

Despite the fact that the count is converted to mitoses per square millimetre, thereby eliminating the effect of the variable size of different HPFs (Ellis and Whitehead 1981), there remain two concerns. Variation between tumours in nuclear size will make the interphase nuclear population, which should be the denominator, also vary between tumours. The other problem is that confining the count to only 10 HPF means that the interphase nuclear population is really quite small. It is readily calculated, using the normal approximation of binomial distribution, that if we want to distinguish between a mitotic index of 0.01 (i.e. 1%) and one of 0.005 (0.5%), we would need to count at least 2235 nuclei and their contained mitoses (Aherne et al. 1977a).

A similar but better method was used in Newcastle upon Tyne in the 1960s for assessing proliferative activity in normal and neoplastic tissues (Simnett and Heppleston 1966). Using an eyepiece, the entire area of the section is measured in terms of viable tissue occupying graticule squares at low power; at high power, representative counts of nuclear density are made per graticule area, allowing calculation of the nuclear count if the magnification factor for the microscope in question is known; then the number of mitoses in the entire section is counted. Counting sufficient blocks, the *mitotic index* is measured from a universe of hundreds of thousands of nuclei. This method has high precision, eliminates subjective bias and measures a genuine kinetic parameter.

Conclusion

It is possible to derive kinetic indices of appropriate precision which do reflect the proliferative status of human tumours, and to use these to develop prognostic indices. It is to be hoped that pathologists will not shrink from the labour necessary to make precise, accurate measurements, rather than beat about looking for short-cuts, which introduce idiosyncratic and statistical inexactitudes.

So where are we? We have a method that is *potentially* simple and reproducible, and is of measurable accuracy and precision. We would advocate a systematic approach in which *mitotic indices*, which reflect tumour heterogeneity, are measured and are encompassed within a confidence interval and from a suitable sample. Only correlation of this variable with tumour recurrence and survival data can answer the doubts and caveats listed above.

References

Aherne WA, Camplejohn RS and Wright NA (1977a) An introduction to cell population kinetics. Edward Arnold, London

Aherne WA, Al-Wiswazy M, Ford D and Kellerer AM (1977b) Assessment of inherent fluctuations of mitotic and labelling indices of human tumours. Br J Cancer 36:577–591

Akerman M, Brandt L, Johnson A and Olsson H (1987) Mitotic activity in non-Hodgkin's lymphoma. Relation to the Kiel classification and to prognosis. Br J Cancer 55:219–223

Appelman HD and Helwig EB (1976) Gastric epithelioid leiomyoma and leiomyosarcoma (leiomyoblastoma). Cancer 38:709–728

Appelman HD and Helwig EB (1977) Cellular leiomyomas of the stomach in 49 patients. Arch Pathol Lab Med 101:373–377

Baak JPA (1990) Mitosis counting in tumors. Hum Pathol 21:683–685

Baak JP, Lindeman J, Overdiep SH and Langley FA (1982) Disagreement of histopathological diagnoses of different pathologists in ovarian tumors – with some theoretical considerations. Eur J Obstet Gynecol Reprod Biol 13:51–55

Baak JP, Langley FA, Talerman A and Delemarre JF (1986) Interpathologist and intrapathologist disagreement in ovarian tumor grading and typing. Anal Quant Cytol Histol 8:354–357

Baak JP, Wisse Brekelmans EC, Uyterlinde AM and Schipper NW (1987) Evaluation of the prognostic value of morphometric features and cellular DNA content in FIGO I ovarian cancer patients. Anal Quant Histol Cytol 9:287–290

Baak JPA, Schipper NW, Wisse-Brekelmans ECM et al. (1988) The prognostic value of morphometrical features and cellular DNA content in cisplatin treated late ovarian cancer patients. Br J Cancer 57:503–508

Baak JPA, van Diest PJ and Ariens ATH, et al. The multicenter morphometric mammary carcinoma project (1989) A nationwide prospective study on reproducibility and prognostic power of routine quantitive assessments in the Netherlands. Pathol Res Pract 185:664–670

Bell DA, Weinstock MA and Scully RE (1988) Peritoneal implants of ovarian serous borderline tumors. Histologic features and prognosis. Cancer 62:2212–2222

Bennett MH, Farrer-Brown G, Henry K and Jelliffe JM (1974) Classification of non Hodgkins lymphoma. Lancet ii:405–406

Bucsema J, Carpenter SE, Rosenshein NB and Woodruff JD (1986) Epithelioid leiomyosarcoma of the uterus. Cancer 57:1192–1196

Burns B, Curry RH and Bell MEA (1979) Morphological features of prognostic significance in uterine smooth muscle tumors: a review of eighty-four cases. Am J Obstet Gynecol 135:109–114

Busch CH and Vasko J (1988) Differential staining of mitoses in tissue sections and cultured cells by a modified methenamine silver method. Lab Invest 59:876–878

Camplejohn RS, Bone G and Aherne WA (1973) Cell proliferation in rectal carcinoma and rectal mucosa. A stathmokinetic study. Eur J Cancer 9:577–581

Christopherson WM, Williamson EO and Gray LA (1972) Leiomyosarcoma of the uterus. Cancer 29:1512–1517

De Potter CR, Praet MM, Slavin RE, Verbeeck P and Roels HJ (1987) Feulgen DNA content and mitotic activity in proliferative breast disease. A comparison with ductal carcinoma in situ. Histopathology 11:1307–1319

De Fusco PA, Gaffey TA, Malkasian GD, Long HJ and Cha SS (1989) Endometrial stromal sarcoma: review of Mayo Clinic experience 1945–1980. Gynecol Oncol 35:8–14

Donhuijsen K (1986) Mitosis counts. Reproducibility and significance in grading of malignancy. Hum Pathol 17:1122–1125

Donhuijsen K (1987) Mitoses in non Hodgkin's lymphoma. Frequency and prognostic relevance. Pathol Res Pract 182:352–357

Donhuijsen K, Schmidt U, Hirche H, van Beunningen D and Budach V (1990) Changes in mitotic rate and cell cycle fractions caused by delayed fixation. Hum Pathol 21:709–714

Duffill M, Wright N and Shuster S (1976) The cell proliferation kinetics of psoriasis examined by three in vivo techniques. Br J Dermatol 94:355–362

Eckstein RP, Paradinas RF and Bagshawe KD (1982) Placental site trophoblastic tumor: a study of four cases requiring hysterectomy including one fatal case. Histopathology 6:211–226

Eckstein RP, Russell P, Friedlander ML, Tattersall MHN and Bradfield A (1985) Metastasising placental site trophoblastic tumor: a case study. Hum Pathol 16:632–636

Edwards JL and Donalson JT (1964) The time of fixation and the mitotic index. Am J Clin Pathol 41:158–162

Ellis PSJ and Whitehead R (1981) Mitosis counting – a need for reappraisal. Hum Pathol 12:3–4

Evans HL (1982) Endometrial stromal sarcoma and poorly differentiated endometrial sarcoma. Cancer 50:2170–2182

Evans HL (1985) Smooth muscle tumors of the gastrointestinal tract. A study of 56 cases followed for a minimum of 10 years. Cancer 56:2242–2250

Evans HL, Butler JJ and Youness EL (1978) Malignant lymphoma, small lymphocytic type. A clinicopathologic study of 84 cases with suggested criteria for intermediate lymphocytic lymphoma. Cancer 41:1440–1455

Friedlander ML (1984) Prognostic variables in epithelial ovarian cancer: A review. Aust NZ Obstet Gynecol 24:256–261

Fu YS, Huang I and Beaudenon S et al. (1988) Correlative study of human papillomavirus DNA, histopathology, and morphometry in cervical condyloma and intraepithelial neoplasia. Int J Gynecol Pathol 7:297–307

Gallup DG and Corday DR (1979) Leiomyosarcoma of the uterus: Case reports and a review. Obstet Gynecol Surv 34:300–312

Gerard-Marchant R, Hamlin I, Lennert K, Rilke F, Stansfeld AG and van Unnik JAM (1974) Classification of Non Hodgkin's lymphoma. Lancet ii:406–408

Gitstein S, Baratz M, David MP and Toaf R (1980) Endometrial stromatosis of uterus. A clinicopathological study of five cases. Gynecol Oncol 9:23–30

Gloor E, Dialdes J, Hurlimann J, Ribolze J and Barrelet L (1983) Placental site trophoblastic tumor of the uterus with metastases and fatal outcome. Clinical and autopsy observations of a case. Am J Surg Pathol 7:483–486

Graem N and Helweg-Larsen K (1979) Mitotic activity and delay in fixation of tumour tissue. Acta Pathol Microbiol Scand Sect A 87:375–378

Griffin NR, Howard MR, Quirke P, O'Brien CJ, Child JA and Bird CC (1988) Prognostic indicators in centroblastic centrocytic lymphoma. J Clin Pathol 41:866–870

Haapasalo H, Collan Y, Atkin NB, Pesonen E and Seppa A (1989a) Prognosis of ovarian carcinomas: prediction by histoquantitative methods. Histopathology 15:167–178

Haapsalo H, Pesonen E and Collan Y (1989b) Volume corrected mitotic index (M/V index) – the standard of mitotic activity in neoplasms. Pathol Res Pract 185:551–554

Haapasalo H, Collan Y, Seppa, Gidlund AL, Atkin NB and Pesonen E (1990) Prognostic value of ovarian carcinoma grading methods – a method comparison study. Histopathology 16:1–7

Hacker NF, Berek JS, Lagasse LD, Nieberg RK and Elashoff RM (1983) Primary cytoreductive surgery for epithelial ovarian cancer. Obstet Gynecol 61:413–420

Hannigan EV and Gomez LG (1979) Uterine leiomyosarcoma. A review of prognostic clinical and pathologic features. Am J Obstet Gynecol 134:557–564

Hart WR and Billman JK (1978) A reassessment of uterine neoplasms originally diagnosed as leiomyosarcomas. Cancer 41:1902–1910

Hemdal I and Frankendal B (1985) Letter to editor. (Comment on paper by Marchesse et al.) Gynecol Oncol 22:379–380

Hermens AF and Barendsen GW (1969) Changes of cell proliferation characteristics in a rat rhabdomyosarcoma before and after X-irradiation. Eur J Cancer 5:173–189

Kempson RL (1976) Mitosis counting II. Hum Pathol 7:482–483

Kempson RL and Bari W (1970) Uterine sarcomas. Classification, diagnosis and prognosis. Hum Pathol 1:331–349

King ME, Dickerson GR and Scully RE (1982) Myxoid leiomyosarcoma of the uterus. A report of six cases. Am J Surg Pathol 6:589–598

Kurman RJ and Norris HJ (1976) Mesenchymal tumors of the uterus. VI. Epithelioid smooth muscle tumors including leiomyoblastoma and clear cell leiomyoma. A clinical and pathological analysis of twenty-six cases. Cancer 37:1853–1865

Kurman RJ, Scully RE and Norris HJ (1976) Trophoblastic pseudotumor of the uterus. Cancer 38:1214–1226

Lathrop JC, Lauchlan S, Nayak R and Ambler M (1988) Clinical characteristics of placental site trophoblastic tumor (PSTT). Gynecol Oncol 31:32–42

Lipponen PK, Eskelinen M, Collan Y, Marin S and Alhara E (1990a) Volume-corrected mitotic index in human pancreatic cancer: relation to histological grade, clinical stage and prognosis. Scand J Gastroenterol 25:548–554

Lipponen PK, Collan Y, Eskelinen M, Pesonen E and Sotarauta M (1990b) Volume-corrected mitotic (M/V index) in human bladder cancer: relation to histological grade (WHO), clinical stage (UICC) and prognosis. Scand J Urol Nephrol 24:39–45

Malkasian GD, Melton J, O'Brien PC and Greene MH (1984) Prognostic significance of histologic classification and grading of epithelial malignancies of the ovary. Am J Obstet Gynecol 149:274–284

Marchesse MJ, Liskow AS, Crum CP, McCaffrey RM and Frick HC (1984) Uterine sarcomas: a clinicopathologic study, 1965–1981. Gynecol Oncol 18:299–312

Morimoto N, Ozawa M, Kato Y and Kuramoto H (1982) Diagnostic value of mitotic activity in endometrial stromal sarcoma. Report of two cases. Acta Cytologica 26:695–704

Norris HJ (1976) Mitosis counting III. Hum Pathol 7:483–484

Norris HJ and Taylor HB (1966) Mesenchymal tumors of the uterus. I. A clinical and pathological study of 53 endometrial stromal tumors. Cancer 19:755–766

Perrone T and Dehner LP (1988) Prognostically favourable "mitotically active" smooth muscle tumors of the uterus. A clinicopathologic study of ten cases. Am J Surg Pathol 12:1–8

Piver MS, Rutledge FN, Copeland L, Webster K, Blumenson L and Suh O (1984) Uterine endolymphatic stromal myosis: a collaborative study. Obstet Gynecol 64:173–178

Quinn CM and Wright NA (1990) The clinical assessment of proliferation and growth in human tumours: evaluation of methods and applications as prognostic variables. J Pathol 160:93–102

Ralfs I, Dawber R, Ryan T, Duffill M and Wright NA (1981) The kinetics of metaphase arrest in human psoriatic epidermis; an examination of optimal experimental conditions for determining the birth rate. Br J Dermatol 104:231–242

Ranchod M and Kempson RL (1977) Smooth muscle tumors of the gastrointestinal tract and retroperitoneum. A pathologic analysis of 100 cases. Cancer 39:255–262

Rappaport H (1966) Tumors of the haematopoietic system. In: Atlas of tumor pathology. Armed Forces Institute of Pathology, Washington DC

Richardson GS, Scully RE, Nikrui N and Nelson JH (1985) Common epithelial cancer of the ovary (first of two parts). N Engl J Med 312:415–424

Rodenburg CJ, Cornelisse CJ, Hermans J and Fleuren GJ (1988) DNA flow cytometry and morphometry as prognostic indicators in advanced ovarian cancer: A step forward in predicting the clinical outcome. Gynecol Oncol 29:176–187

Ronan SG, Han MC and Das Gupta TK (1988) Histologic indicators in cutaneous malignant melanoma. Semin Oncol 15:558–565

Sadler DW and Coghill SB (1989) Histopathologists, malignancies and undefined high power fields. Lancet i:785–786

Saksela E, Lampinen V and Procope BJ (1974) Malignant mesenchymal tumors of the uterine corpus. Am J Obstet Gynecol 120:452–460

Scully RE and Young RH (1981) Trophoblastic pseudotumor. A reappraisal. Am J Surg Pathol 5:75–76

Silverberg SG (1971) Leiomyosarcoma of the uterus. A clinicopathologic study. Obstet Gynecol 38:613–628

Silverberg SG (1976) Reproducibility of the mitosis count in the histologic diagnosis of smooth muscle tumors of the uterus. Hum Pathol 7:451–454

Simnett JD and Heppleston AG (1966) Cell renewal in the mouse lung. The influence of sex, strain and age. Lab Invest 15:1793–1801

Sorbe B, Frankendal B and Veress B (1982) Importance of histologic grading in the prognosis of epithelial ovarian carcinoma. Obstet Gynecol 59:576–582

Stansfeld AG, Diebold J and Noel H, et al. (1988) Updated Kiel classification for lymphomas. Lancet i:292–293

Start RD, Cross SS and Smith JFH (1990) Effect of delayed fixation in the assessment of tumour cell proliferation. J Pathol 161:183–184

Swerdlow SH, Habeshaw JA, Dhaliwall HS, Lister TA and Stansfeld AG (1983) Centrocytic lymphoma: a distinct clinicopathologic and clinical entity. Am J Pathol 113:181–197

Taina E, Maenpaa J, Erkkola R, Ikkala J, Soderstrom O and Viitanen A (1989) Endometrial stromal sarcoma: a report of nine cases. Gynecol Oncol 32:156–162

Tannock IF (1968) The relation between cell proliferation and the vascular system in a transplanted mouse mammary tumour. Br J Cancer 22:258–273

Tannock IF (1970) Population kinetics of carcinoma cells, capillary endothelial cells and fibroblasts in a transplanted mouse mammary tumor. Cancer Res 30:2470–2476

Taylor HB and Norris HJ (1966) Mesenchymal tumors of the uterus. IV. Diagnosis and prognosis of leiomyosarcomas. Arch Pathol 82:40–44

Taylor JF, Iversen OH and Bjerknes R (1977) Growth kinetics of Kaposi's sarcoma. Br J Cancer 35:470–478

Thatcher SS and Woodruff JD (1982) Uterine stromatosis: a report of 33 cases. Obstet Gynecol 59:428–434

The Non-Hodgkin's lymphoma pathologic classification project: National Cancer Institute sponsored study of classifications of Non-Hodgkin's lymphoma (1982) Summary and description of a working formulation for clinical usage. Cancer 49:2112–2135

Tiltman AJ (1988) Muscle tumor mitoses. Am J Surg Pathol 12:967–968

Twiggs LB, Okagaki T, Phillips GL, Stroemer JR and Adcock LL (1981) Trophoblastic pseudotumor-evidence of malignant disease potential. Gynecol Oncol 12:238–248

Wright NA (1984) Cell proliferation in health and disease. In: Anthony PP, MacSween RNM (eds) Recent advances in histopathology. Churchill Livingstone, Edinburgh, pp 17–33

Wright NA and Alison M (1984) The biology of epithelial cell populations. Clarendon Press, Oxford

Wright NA and Appleton DR (1980) The metaphase arrest technique. A critical review. Cell Tissue Kinet 13:643–663

Wright NA, Britton DC, Bone G and Appleton DR (1977) An in vivo stathmokinetic study of cell proliferation in human gastric carcinoma and gastric mucosa. Cell Tissue Kinet 10:429–438

Wright NA, Appleton DR, Marks J and Watson AJ (1979) Cytokinetic studies of crypts in convoluted human small intestinal mucosa. J Clin Pathol 32:462–470

Yoonessi M and Hart WR (1977) Endometrial stromal sarcomas. Cancer 40:898–906

Young RH and Scully RE (1984) Placental site trophoblastic tumor: current status. Clin Obstet Gynecol 27:248–258

Zaloudek C and Norris HJ (1987) Mesenchymal tumors of the uterus. In: Kurman RJ (ed). Blaustein's pathology of the female genital tract. Springer Verlag, New York, pp 373–408

6 Flow Cytometry

R.S. CAMPLEJOHN and J.C. MACARTNEY

Introduction

Development and Properties of Flow Cytometers

Flow cytometry has developed over the past 30 years from initial attempts to count and size particles. The first cell sorter was described in 1965, and multiparameter machines measuring two fluorescence wavelengths were described around 1970. An excellent summary of the historical development of flow cytometers is given by Melamed et al. (1990a).

Currently two main categories of flow cytometer are commercially available. Firstly, there are simple to use bench-top machines which typically can measure five simultaneous parameters on single cells or nuclei as they pass through the cytometer. Three of these parameters are related to the fluorescence characteristics and two to the light scattering properties of the cells. The second major category of flow cytometer is that of large cell sorters. These machines not only measure five or more parameters on particles passing through them but they can also physically sort particles with a desired set of properties into separate containers for further study. Many cell sorters have the ability to use two lasers for excitation of cellular fluorescence, which allows a wider variety of fluorochromes to be used. Cell sorting will not be discussed further in this chapter, but two good books are available which describe in detail the principles and methods relating to cell sorters and flow cytometers (Melamed et al. 1990b; Ormerod 1990).

The basic principles of flow cytometry are simple (Fig. 6.1). Particles such as cells or nuclei pass singly through a focused light beam produced most often by a laser. This light beam, of a particular, desired wavelength, excites fluorescent dyes used to stain cellular constituents of interest such as DNA. The fluorescence from one or more dyes is collected, split into its different components (if more than one dye is involved) by a series of mirrors and lenses, and measured by detectors. Signals from the detectors are then processed electronically into a form suitable for storage and analysis on a computer.

Techniques are now available to measure flow cytometrically a wide variety of cellular constituents and properties. For example, cellular DNA, RNA and protein content can be measured as, with the aid of monoclonal antibodies, can the presence of specific proteins and other molecules in individual cells.

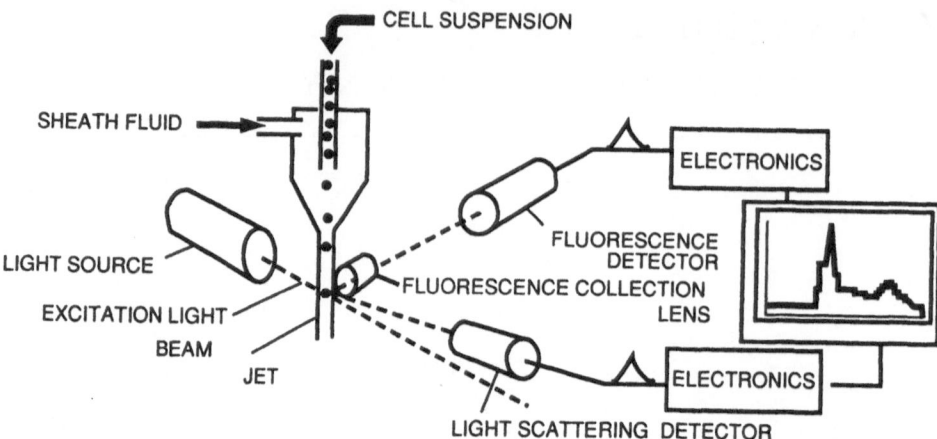

Figure 6.1. Schematic representation of a flow cytometer, simplified to show one fluorescence and one light scatter detector (redrawn from Steen 1990).

Intracellular pH, changes in intracellular calcium levels and various aspects of cellular metabolism can be assessed. Furthermore, two or more of these factors can be measured simultaneously on individual cells, and this enables interesting functional studies to be performed (Melamed et al. 1990b).

General Advantages and Disadvantages of Flow Cytometry

Flow cytometry has the advantages of speed and statistical precision. Typically 10 000–100 000 cells or nuclei can be scanned in a few minutes or less. Given suitable staining conditions, measurements of fluorescence in individual particles are related quantitatively to the amount of the stained substance present. Multiple parameters can be measured simultaneously on individual cells: this can be a useful attribute of the technique.

No technique is without its disadvantages, however, and flow cytometry is no exception. Amongst the disadvantages of flow cytometry is that a relatively expensive machine is required. In addition, when studying solid tissues, the need to disaggregate the tissue into a suspension of single cells or nuclei can be a problem. Some solid tissues are difficult to disaggregate, and in all cases tissue morphology is lost.

DNA Flow Cytometry

Introduction

The largest body of clinically relevant flow cytometric studies related to proliferative activity involves the measurement of DNA content. The earliest flow cytometrically measured DNA histograms, which showed clearly defined G_1, S and G_2+M phases of the cell cycle seem to have been produced by Van Dilla

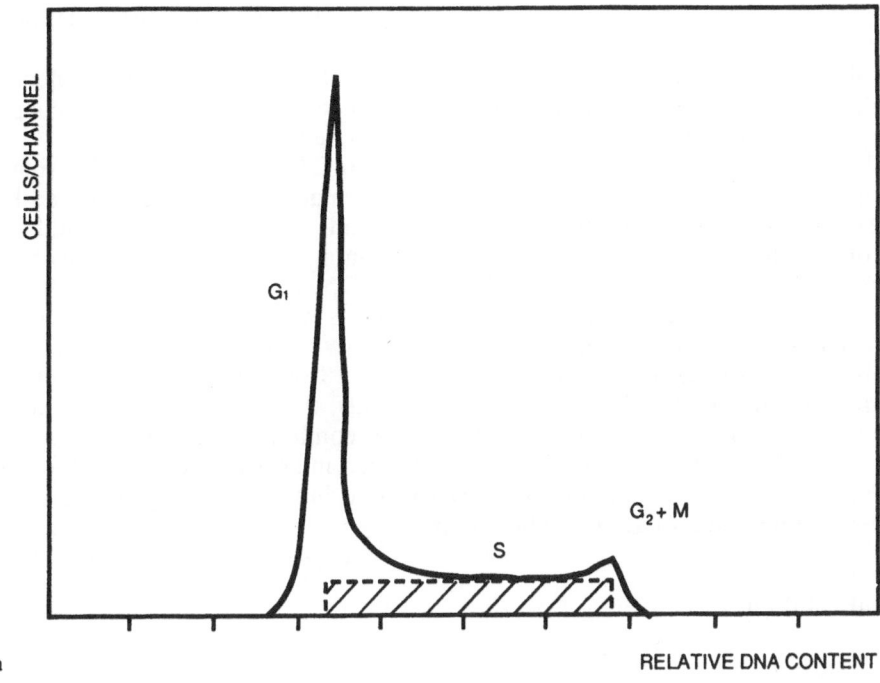

a

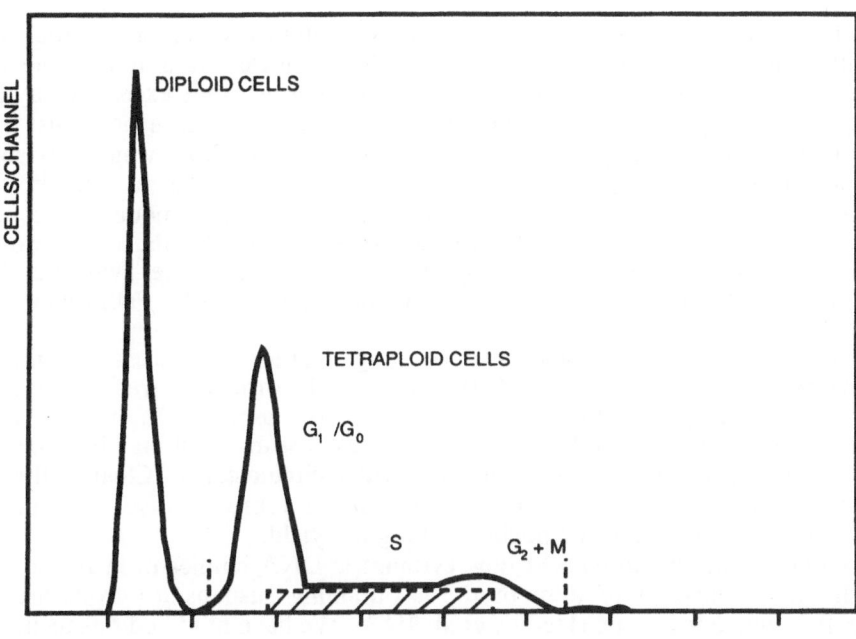

b

Figure 6.2. a A diploid histogram. The quality of such histograms is usually expressed in terms of the CV of the G_1 peak; the smaller the CV the better the quality. SPF is calculated by the method of Baisch et al. (1975), which involves fitting a rectangular area to represent the S phase. **b** A DNA histogram for a tetraploid tumour, in which the diploid G_1 peak represents largely non-proliferative normal cells. The tumour G_1 peak has twice the normal amount of DNA, and thus has a DNA index (DI) of 2.0. The SPF is calculated by a modified version of the Baisch technique (Camplejohn et al. 1989).

et al. (1969). Since then a vast number of studies have been published, in which DNA content of clinical material has been measured. In this chapter discussion will be restricted to topics related to neoplasia.

Two main parameters can be measured from DNA histograms. The first of these identifies the presence of cells with abnormal amounts of DNA (see Fig. 6.2), so-called DNA aneuploid cells. Early studies using both static and flow cytometry tended to show that tumours containing DNA aneuploid cells had a worse prognosis than those containing only DNA diploid cells. The second major parameter gleaned from DNA histograms is a crude index of proliferative activity. In general, where proliferative activity is high there will be many cells in the S and G_2+M phases of the cell cycle. Conversely when proliferative activity is low there will be few such cells. Thus, the percentage of cells in the S phase (the S-phase fraction or SPF) or the combined percentage of cells in the S and G_2+M phases are often used as crude indices of proliferative activity (Fig. 6.2). The same limitations apply to these indices as apply to the use of tritiated thymidine or BUdR-labelling index.

Methodology

Preparation of Cell/Nuclear Suspensions

There are many published papers describing methods for disaggregating fresh solid tissues into suspensions of either cells or nuclei. It is impossible in a chapter of this length to go into detail about such methods except to say that they can be subdivided into a number of categories. For cell suspensions, there are two main types of method. Firstly, there are the methods which rely solely on mechanical dissaggregation (in our experience these tend to work best on tissues which are relatively easy to dissociate such as lymph nodes). Secondly, there are methods which involve the use of enzymes. For the production of nuclear suspensions various techniques using detergents are available. Disaggregation methods for fresh tissue have been discussed by Pallavicini et al. (1990).

Attempts have been made, with varying degrees of success, to develop disaggregation methods that will work on a wide variety of fresh tissues and tumours (Ensley et al. 1987; Vindelov and Christensen 1990). In our experience no method works well on all types of fresh tissue and it may be necessary to modify a given technique to fit a particular circumstance. Clearly, the aim of any disaggregation method is to produce a representative, high-quality suspension of single cells or nuclei with a good yield.

A major stimulus to clinical flow cytometric DNA studies occurred in 1983 with the publication of a method for obtaining suspensions from archival paraffin-embedded tissue (Hedley et al. 1983). We have modified this technique and automated part of it (Camplejohn et al. 1989). The method we use is summarised in Figure 6.3. The use of such material has a number of advantages for clinical studies. Firstly, large retrospective series of patients for whom clinical follow-up is already available can be investigated. This is particularly useful for breast cancer, for example, where 10 or more years follow-up is preferable before conclusions relating to survival are made. Moreover, if one wishes to study rare lesions the use of archival material enables an adequate

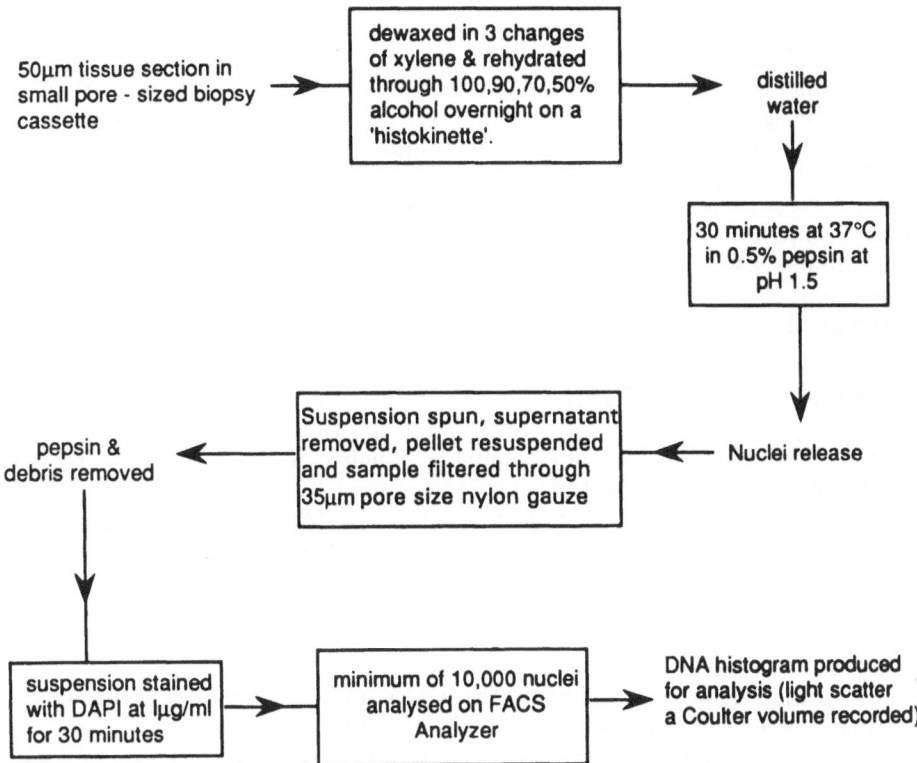

Figure 6.3. Schematic representation of method for producing DNA histograms from paraffin-embedded material.

series of patients to be collected. DNA flow cytometric studies from paraffin are also logistically easy for large multi-centre studies where the flow cytometry is to be performed centrally.

What then are the disadvantages of using paraffin-embedded material? One of the most common disadvantages quoted is that the quality of the histograms produced is poorer, and this was noted by Hedley et al. (1983) in their original study. We have not found this poorer quality in a comparison of fresh and paraffin-embedded lymph node biopsies (Camplejohn et al. 1989), but most workers who have done comparisons do find that paraffin-embedded tissue yields results of poorer quality (Kallioniemi 1988).

Section thickness affects the quality of the results achieved (Stephenson et al. 1986) with >50 μm being optimal. However, as has been pointed out (Camplejohn and Macartney 1986) this is not the only factor involved. Often poor-quality results are associated with large amounts of debris in the samples. This may result from inappropriate fixation and processing of the tissue prior to performance of the flow cytometry. Bouin's fixative and various types of mercuric fixatives have in general yielded poor results (Hedley 1989). In our hands, the best fixation protocol was neutral buffered formalin at 4°C overnight. Given a suitable fixation and processing protocol good quality results are achievable from most paraffin-embedded samples (Gillet et al. 1990; our unpublished data).

In general, in studies comparing DNA index, the results from fresh and paraffin-embedded tissue have shown good agreement (Camplejohn et al. 1989; Hedley et al. 1983; Kallioniemi 1988; McIntire et al. 1987; Schutte et al. 1985). The agreement for SPF is also fairly good but with more variability in some studies. Camplejohn et al. (1989) found good accord between SPF in 62 lymph node samples analysed fresh and paraffin-embedded, while McIntire et al. (1987) in a similar but smaller study found a poorer correlation. Kallioniemi (1988) compared SPF in 50 frozen and paraffin-embedded breast cancer samples and found reasonable agreement. The important question, which we will address later, is whether the SPF and the DNA index (DI) however derived, have any clinical value.

Staining and Flow Cytometric Measurement

A variety of fluorochromes are available which can be used to stain DNA quantitatively. A detailed description of the properties of these dyes is given in three consecutive chapters of the book edited by Melamed and colleagues in 1990 (Waggoner 1990; Crissman and Steinkamp 1990; Latt and Langlois 1990). While not discussing the physico-chemical properties of these dyes in the present chapter, we feel that one or two points should be made.

With all these dyes, it is necessary to get a stain : cell concentration ratio that achieves stoichiometric staining. This involves the dye being present in modest excess so that all available sites on the DNA are bound by dye. Only in this way can quantitative measurement of DNA content be achieved.

A number of DNA dyes, such as PI, require cells or nuclei to be treated with RNAse, as they also stain double-stranded RNA. This is one of the reasons we have used DAPI as the DNA fluorochrome for routine flow cytometry of paraffin sections. This dye binds only to DNA, thus RNAse is not required, which reduces the cost per sample significantly. However, McIntire et al. (1987) have suggested that values of SPF estimated from samples stained with DAPI are different to those obtained with PI. We were concerned at this possibility and made a larger, more detailed comparison of PI and DAPI staining on a series of lymph node biopsies (Camplejohn et al. 1989). We found good agreement between results in terms of DI and SPF using the two dyes on both fresh and paraffin-embedded material.

Data Analysis

Perhaps the most complex and vexatious aspect of DNA flow cytometry is data analysis. Some attempts have been and are being made, for example, by the International Society for Analytical Cytometry to standardise procedures and nomenclature (Hiddemann et al. 1984). However, many different methods are used to analyse data, often with the aid of a variety of computer algorithms. No attempt will be made here to review the literature on this topic, but reference can be made to Dean (1990) and Ormerod (1990). In the present chapter we will mainly describe what is done in our laboratory, with some explanation as to why we use the methods described.

The measurement of DNA ploidy will be considered first. The basic principles by which DNA aneuploidy is recognised and by which DNA index is calculated are widely accepted (Fig. 6.2). With fresh tissue samples external DNA standards can be used as a means of confirming the DNA ploidy status of a given DNA peak in a test sample. The most accepted way of doing this is to use human peripheral blood cells (Hiddemann et al. 1984). Such external DNA standards cannot be used with paraffin-embedded material, as was pointed out in the original paper by Hedley et al. (1983) and has been confirmed many times since (see, for example, Schutte et al. 1985). The reason for this is the variability in DNA staining intensity in nuclei from different paraffin blocks due to differences in fixation and processing. Attempts have been made to use, as diploid standards, nuclei taken from a normal part of the same specimen as the tumour nuclei (Schutte et al. 1985). However, such normal tissue is not always available, and even when it is doubt has been cast on the reliability of this practice (Price and Herman 1990). It is safest to avoid the use of all types of external standard for DNA determinations from paraffin-embedded material. With DNA histograms from such material, the convention is that where two G_1 peaks are seen the left-hand peak is considered to be DNA diploid and the DI of the other peak is calculated accordingly. This should not lead to the failure to recognise DNA aneuploidy, as all samples contain reference DNA-diploid cells (lymphocytes, endothelial cells, etc.), but it will lead to a small percentage of samples being wrongly classified as hyperdiploid when they are really hypodiploid. The clinical significance of such an incorrect classification is unknown but in most instances is likely to be minor.

Two common problems in assessing DNA ploidy status are the difficulties in recognising either small DNA-aneuploid stem lines or even worse, small tetraploid stem lines. These problems can be more easily dealt with if other flow cytometric parameters, which are freely available on most machines, such as 90° light scatter and forward light scatter (or Coulter volume) are used (see Fig. 6.4).

For the estimation of SPF (Fig. 6.2), we use the method of Baisch et al. (1975) for DNA-diploid histograms and a modification of this method for DNA-aneuploid histograms (Camplejohn et al. 1989). We chose this method of calculation for a variety of reasons. Firstly, it is simple and requires only a hand-held calculator. Thus, it can be applied anywhere and does not require any specific computer hardware or software. Secondly, in a study of a series of cell lines it has given good agreement with the BUdR labelling index (unpublished data).

If one looks at the agreement between the SPF and the tritiated thymidine or BUdR labelling indices, there is in general fairly good agreement between the two methods (Braylan et al. 1980; Costa et al. 1981; Linden et al. 1980; McDivitt et al. 1986; Silvestrini et al. 1988). However, some of the studies did find discrepancies and the two parameters, namely SPF and labelling index, cannot be assumed to be equivalent, although they are clearly related.

Summary

Clearly there are technical differences between the many laboratories around the world in which DNA flow cytometry is performed. Thus, it is not possible to assume for example, that an SPF of 5% is equivalent in all laboratories. However, as we will discuss shortly, trends shown by different laboratories may

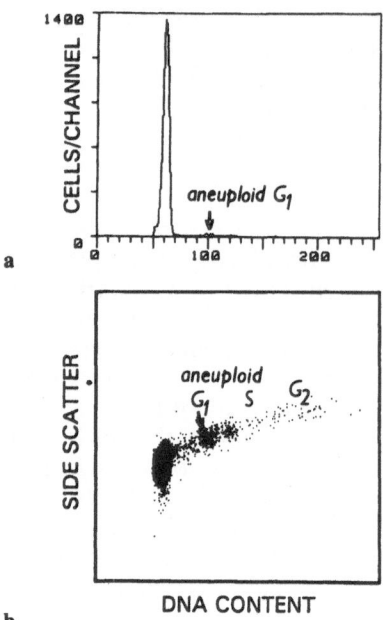

a

b

DNA CONTENT

Figure 6.4. a A DNA histogram prepared from a paraffin-embedded breast cancer sample. A barely discernible aneuploid peak is visible. By combining DNA measurement with side scatter, a clearly discernible collection of aneuploid G_1 cells can be seen with corresponding S and G_2 + M phases (**b**).

be similar even if the actual values achieved are not. There is indeed evidence that if a similar technique is used, then comparable results are achieved on different sites (Camplejohn and Macartney 1986; Coon et al. 1988).

Clinical Value

Criteria for Useful Clinical Studies

Virtually all clinical DNA flow cytometric studies are performed in situations where routine histopathological techniques are also applied. Standard histopathological techniques are well established and it seems unlikely that flow cytometry will to any extent replace such methods. Thus, if flow cytometry is to be performed as well as routine histopathology, it must add something to it. Further, for routine clinical application the additional information gleaned from flow cytometry should be of use in patient management. It should help in making choices as to how to best manage individual patients. In recent years, we have attempted to focus our clinical DNA flow cytometric work in areas where the above criteria might apply.

Diagnostic Applications

For the purposes of this chapter the term "diagnosis" will be used in the narrow sense of distinguishing between neoplastic and normal cells.

Despite the fact that much of the original impetus to develop flow cytometers came from a desire to automate *diagnostic* cytology, the majority of studies now relate to the *prognostic* use of the technique. A good summary of flow cytometry in clinical cytology is given by Melamed and Staiano-Coico (1990). The two main areas of application are uterine cervical and bladder cytology. A series of early studies suggested a possible role for flow cytometry in cervical cytology, but interest in this area has waned somewhat recently. DNA as a parameter is important but alone it is inadequate for identifying samples containing abnormal cells (Melamed and Staiano-Coico 1990): too many positive samples are misclassified as normal. Multiparameter flow measurements involving DNA, combined with light scatter and antibody staining, may help in the future.

Progress in applying flow cytometry in the field of bladder cytology has been more promising. Some studies claim that DNA measurements are of value alone (Koss et al. 1989), but there may be problems with heavy contamination by non-malignant cells. A series of papers has been published by the group from the Sloan–Kettering Cancer Center in New York, on the use of acridine orange staining to measure DNA and RNA simultaneously. The same group has also done many technical investigations into, for example, the difference between voided urine and bladder washings as a source of cells and the difference between samples obtained from men and women (for a more complete bibliography, see Badalament et al. 1987; Hermansen et al. 1989). While the results of this series of studies look promising in terms of a role for flow cytometry as an aid in diagnosing bladder cancer, the acridine orange technique is not easy to perform or standardise.

Flow cytometry has been applied in other areas such as pleural and ascitic fluids and bone marrow (Melamed and Staiano-Coico 1990). An interesting diagnostic application of DNA flow cytometry relates to the diagnosis of molar pregnancy (Hemming et al. 1987). As yet, however, none of the diagnostic flow cytometric techniques are widely used and their potential for routine clinical application remains uncertain.

Application in Prognosis

The literature on clinical studies of DNA flow cytometry, in which the aim has been to compare DNA results with other known prognostic factors or to relate the flow results directly to patients' survival, is vast. It is therefore impossible in the space available here, to try and review the whole field or even to deal superficially with all the major types of cancer. However, a number of review articles have been published, which would give the interested reader a starting point in obtaining data on a wide variety of tumour types (Barlogie 1984; Friedlander et al. 1984; Hedley 1989; Merkel and McGuire 1990; Raber and Barlogie 1990). In this short section, we will use two major types of cancer, namely non-Hodgkin's lymphoma (NHL) and breast cancer, as examples of the type of clinical data available. In both, clinically interesting results have been achieved (see also Chap. 10). Both diseases have been well studied, to the extent that even restricting the discussion to the two of them will require a fairly superficial overview.

Non-Hodgkin's Lymphoma. The discussion in this section is based on a recently published feature article (Macartney and Camplejohn 1990), which can be referred to for a more complete bibliography.

DNA aneuploidy is more common in high-grade NHL, but the presence of DNA-aneuploid cells has little prognostic significance. Three studies in high-grade NHL did find that DNA aneuploidy had an adverse effect on survival, but one study found the opposite (Macartney and Camplejohn 1990). In none of the studies was an effect demonstrable after multivariate analysis.

Due to inconsistent flow cytometric methods, inconsistent pathological grading and uneven statistical analysis, it is difficult to compare the published reports of proliferative activity in NHL (Macartney and Camplejohn 1990). In general, proliferative activity does correlate with histological grade. There is, however, in all published studies, substantial heterogeneity in proliferative activity within grades or histological subtypes of NHL, as well as between them. This limits the value of proliferative activity as a means of classifying or grading NHL. However, this very heterogeneity within supposedly homogeneous histological subtypes of NHL may make this measurement of prognostic value within a given subtype. As discussed earlier under criteria for clinical studies, any prognostic information must be additional to that obtained from standard histopathology, otherwise DNA flow cytometry is little more than an expensive exercise in tumour grading.

There have been three studies that reported a worse prognosis for low-grade follicular NHL with high proliferative activity (SPF > 5%), either in terms of overall survival (Rehn et al. 1990; Macartney et al. 1991) or increased probability of high-grade transformation (Macartney et al. 1986). (Data from Macartney et al. (1991) is given in Fig. 6.5.) The two most recent studies (Rehn et al. 1990; Macartney et al. 1991) both show that B symptoms and high proliferative activity were the strongest predictors of survival in multivariate analyses. Thus, low-grade follicular NHL may be the sort of situation where DNA flow cytometry can aid patient management. There seems to be a case for treating follicular NHL with high proliferative activity with more aggressive therapeutic regimes.

The situation in high-grade lymphoma is more complex (see Macartney and Camplejohn 1990), but there appears to be an adverse effect of high proliferative activity on survival, at least in the short term (see Chaps. 10 and 11).

Breast Cancer. A large number of flow cytometric studies have been performed on this disease. Many of the earlier studies were restricted to DNA ploidy and many examined only the association between flow cytometric parameters and other pretreatment variables, not survival. To summarise the findings of these studies (for bibliography see Merkel and McGuire 1990; O'Reilly et al. 1990a), DNA aneuploidy is more common in high-grade tumours in virtually all published studies. The correlation of ploidy with other clinico-pathological variables is less certain. Some groups reported an association between the frequency of DNA aneuploidy and tumour size, while other groups did not find this, and a similar story is found for nodal status and DNA ploidy (see Merkel and McGuire 1990). The relationship between steroid receptor status and DNA ploidy is also uncertain. Merkel and McGuire (1990) found eleven studies (involving 2553 patients) in which DNA aneuploidy occurred more commonly in estrogen receptor negative tumours, and six studies (involving 752 patients) in which no relationship or a non-significant trend was found. Our studies (O'Reilly et al. 1990a) would fit into this second group. The impression from reviewing the literature is that many of these parameters may be interrelated but that the correlations are weak.

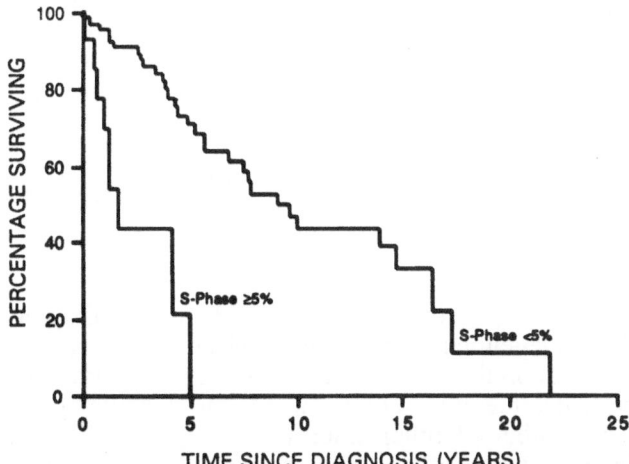

Figure 6.5. Survival curves for follicular lymphomas with SPF below ($n = 67$) and above ($n = 16$) 5%. SPF retained its significance when treated as a continuous variable.

As regards SPF, there is little evidence that it correlates with tumour size or nodal involvement, but many studies find a correlation with steroid receptor status (Merkel and McGuire 1990). SPF does correlate, usually strongly, with tumour grade in all studies in which this association was investigated.

In terms of predicting survival, two distinct questions should be kept in mind. Firstly, do DNA ploidy and/or SPF predict for survival in meaningful groups of breast cancer patients? Secondly, is any such predictive power independent of other known clinico-pathological factors? As regards DNA ploidy, there is some disagreement for both node-negative and node-positive patients in answering these questions. Many of the earlier studies are reviewed by Merkel and McGuire (1990). For node-negative disease a number of recent reports illustrate the disagreement. In a recent study, we failed to find a statistically significant association between DNA ploidy and survival (O'Reilly et al. 1990b). In a group of early node-negative patients Toikkanen et al. (1990) found an association between patients with DI greater than or less than 1.2 and survival if univariate analysis was used, but this predictive ability was lost if multivariate analysis was used. In contrast, Clark et al. (1989) found that DNA ploidy was predictive for relapse-free survival even in a multivariate analysis, although tumour grade was not a factor included in this analysis. A similar lack of agreement exists concerning the relationship between DNA ploidy and survival in node-positive disease, with some studies finding no association, others an association on univariate analysis only and some a statistically independent relationship (see Merkel and McGuire 1990; Toikkanen et al. 1989; O'Reilly et al. 1990a). In summary, it seems that most studies find, using univariate analysis at least, an association between DNA ploidy and such as steroid-receptor status, and tumour size and age, were included but not performed this predictive power is in most cases lost.

In the past 5 years, more studies have looked at SPF as well as DNA ploidy. In both node-negative and node-positive patients, tumour SPF was found in nearly all published studies to correlate, usually strongly, with survival (see Merkel and McGuire 1990 for a summary of earlier studies; Clark et al. 1989; Kallioniemi et al. 1988; Muss et al. 1989; O'Reilly et al. 1990a, b; Toikkanen

Table 6.1. A summary of relapse-free survival (RFS) versus tumour size and SPF in a group of 150 node-negative breast cancer patients (data from O'Reilly et al. 1990b)

Group	Tumour size + SPF	No.	5-year RFS (%)
1	<1 cm	33	96
2	>1 cm + low SPF	76	72
3	>1 cm + high SPF	41	52

et al. 1989, 1990; Uyterlinde et al. 1990). Nevertheless, there are disagreements as to whether SPF has independent prognostic significance, particularly, if tumour grade is included in the multivariate analysis. At least one study has found SPF to be of independent value in a multivariate analysis including grade (Toikkanen et al. 1989). A number of other studies have reported it to be of independent prognostic value when other clinico-pathological parameters, such as steroid-receptor status, and tumour size and age, were included but not tumour grade (Clark et al. 1989; Kallioneimi et al. 1988; Muss et al. 1989; O'Reilly et al. 1990a, b). In general, the impression is that SPF is probably a stronger prognostic indicator than DNA ploidy in breast cancer. That is certainly the finding of our own studies (O'Reilly et al. 1990a, b).

As an example of studies focused on a specific clinical question, we shall briefly describe an investigation into node-negative breast cancer. Although survival overall is better for patients with no nodes involved than for those with involved nodes, there is a sub-group of node-negative patients who do poorly. It would not generally be acceptable to give all node-negative patients toxic adjuvant therapy, but it might be worth giving such treatment to suitable patients with a poor prognosis. How to identify such patients is thus a clinically relevant question. We examined whether DNA flow cytometry might be a useful way to do this (O'Reilly et al. 1990b). The results suggest that by combining tumour size and SPF, a poor prognosis sub-group of node-negative patients can be identified (see Table 6.1).

Multiparameter Flow Cytometry

As was described earlier, it is possible to analyse several fluorescent probes simultaneously by means of flow cytometry. Multiparametric flow studies offer exciting possibilities in a wide area of cell and molecular biology (see Melamed et al. 1990b and Ormerod 1990 for more details). But do multiparametric measurements have a role in clinical studies?

We have already described the value of light scatter/volume measurements to help in the analysis of DNA histograms (Fig. 6.4). It may be possible to extend this application of light scatter to distinguish normal from malignant nuclei from paraffin sections, even when both are DNA diploid (Ormerod and Imrie 1991). Admixture of normal and malignant cells is a serious problem if one is interested in calculating SPF, particularly, for DNA-diploid tumours.

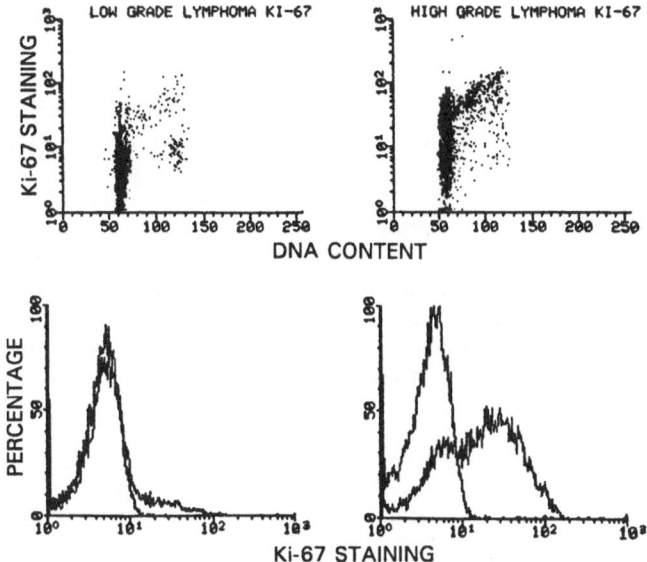

Figure 6.6. *Top two panels:* DNA content (*abscissa*, linear scale) plotted against Ki-67 staining (log scale). *Lower two panels:* Ki-67 staining plotted as histograms with a negative control overlaid. *Left-hand panels* show results for a low-grade lymphoma with an SPF of 2% and a Ki-67-positive cell population constituting 6% of the total. *Right-hand panels* illustrate the equivalent results for a high-grade lymphoma (SPF 20% and 60% Ki-67-positive cells). In both cases it can be seen that, effectively, all S-phase cells are positively stained.

With paraffin-embedded material, there is no known way of distinguishing normal diploid from malignant diploid cells, unless the method of Ormerod and Imrie (1991) proves to have wide application. For most DNA-aneuploid tumours, the SPF of the aneuploid cells can be calculated from a part of the histogram not contaminated by DNA diploid cells (Fig. 6.2). If one is working with fresh tissue, combining surface or cytoplasmic markers with DNA may aid in producing DNA histograms relatively uncontaminated by normal cells. This approach has been applied with some success in a variety of tumour types, for example NHL (Braylan et al. 1984) and colorectal adenocarcinoma (Crissman et al. 1989). Thus, multiparametric flow cytometry, be it with light scatter and/or markers, may help to produce better quality tumour DNA profiles by excluding normal cells from the analysis.

Multiparameter flow cytometry could offer much more to clinical studies than simply helping to distinguish between normal and malignant cells. It is impossible in the space available here to discuss many of the areas of promise, but we have already mentioned the combined measurement of RNA and DNA (see also Darzynkiewicz and Kapuscinski 1990). A further multiparametric method involving BUdR is described in detail by Wilson and McNally in Chapter 7. This technique can give kinetic information as well as measuring static indices of proliferation.

There are also now available antibodies to many proliferation-related proteins, most of which are found in the nucleus of proliferating cells. Many of these have been measured flow cytometrically, usually combined with DNA content, and it may be that such measurements could be of clinical value in the future. Amongst the proliferation-related markers already measured flow cytometrically on clinical material are the c-*myc* oncogene product (Watson et al. 1985, 1986), Ki-67 (Drach et al. 1989), PCNA (Kurki et al. 1988) and p53 (Remvikos et al. 1990). An example of such a marker measured together with DNA is given in Fig. 6.6. (data from our own studies with Ki-67 in NHL). A more detailed summary of proliferation-related markers measured flow cytometrically is given by Darzynkiewicz (1990). Some of these markers can be detected in nuclei extracted from paraffin-embedded sections but caution should be used in interpreting such results. Changes in specific protein levels due to degradation and/or redistribution during fixation and processing could lead to spurious findings (Lincoln and Bauer 1989). Nevertheless, the measurement of such proliferation-related markers on fresh, and possibly paraffin-embedded tissue, promises to be an area of great interest.

Conclusion

DNA flow cytometry certainly seems to be of clinical value in a variety of tumours, including NHL and breast cancer. Whether it is the best or most cost-effective means of assessing proliferative activity in clinical material remains to be established. What is clear is that it is in routine use in, for example, breast cancer. In the USA the majority of breast cancer patients have DNA flow cytometry performed as part of their routine workup (McGuire personal communication). A single laboratory in California runs between 30 000 and 40 000 breast tumour samples per year.

Ideally, it would be best to standardise pathological classification of disease types, flow cytometric and staining techniques and the methods of data analysis, as outlined by Macartney and Camplejohn (1990) for NHL. However, in practice, this is hard to achieve. Many technical differences have historical roots but others are the result of genuinely held differences of opinion. One of the many examples of this relates to the use of paraffin-embedded or fresh tissue for prospective studies. We feel that with more careful fixation of samples (see earlier), the quality of results from paraffin-embedded tissue can be brought closer to that obtained from fresh tissue. While there is no doubt that the use of fresh tissue samples is ideal, in our hospital situation there are often practical restraints upon its use. We are unable, from the group of hospitals we work with, to obtain fresh tissue samples from even the majority, much less all, breast cancer cases. When performing flow cytometry centrally for multiple centres, there are serious logistical problems in obtaining and transporting fresh tissue samples so that they arrive in a satisfactory state. Further, particularly with small lesions, pathologists are often reluctant to release pieces of fresh tissue from samples taken for diagnostic purposes. Thus, we feel that the use of paraffin-embedded tissue is a reasonable way to proceed for some routine clinical applications of DNA flow cytometry. In contrast, Vindelov and Christensen (1990) in a detailed strategy for clinical DNA studies, reject the use of paraffin-embedded tissue on the grounds of the inferior quality of results obtained.

Such differences will be difficult to reconcile. Although this does not prevent the application of DNA flow cytometric data to clinical management, it does necessitate caution in comparing results from different centres, unless they use similar methods. The general trend of results will hopefully be similar in different centres but the precise values of SPF, for example, may not be. Certainly some steps towards standardisation could be taken, notwithstanding the difficulties listed above. For example, all current and future studies should look at the clinical value of flow cytometrically derived data in a meaningful multivariate analysis, which includes all known, relevant clinico-pathological parameters. In this way, the significance of the results obtained should be far easier to assess.

The promise of clinically useful multiparameter flow studies is largely for the future. The prospects certainly look interesting.

References

Badalament RA, Hermansen DK and Kimmel M, et al. (1987) The sensitivity of bladder wash flow cytometry, bladder wash cytology, and voided cytology in the detection of bladder carcinoma. Cancer 60:1423–1427

Baisch H, Gohde W and Linden WA (1975) Analysis of PCP-data to determine the fraction of cells in the various phases of cell cycle. Radiat Environ Biophys 12:31–39

Barlogie B (1984) Abnormal cellular DNA content as a marker of neoplasia. Eur J Cancer Clin Oncol 20:1123–1125

Braylan RC, Diamond LW, Powell ML and Harty-Golder B (1980) Percentage of cells in the S phase of the cell cycle in human lymphoma determined by flow cytometry. Cytometry 1:171–174

Braylan RC, Benson NA and Nourse VA (1984) Cellular DNA of human neoplastic B-cells measured by flow cytometry. Cancer Res 44:5010–5016

Camplejohn RS and Macartney JC (1986) Comments on "effect of section thickness on quality of flow cytometric DNA content determinations in paraffin-embedded tissues". Cytometry 7:612–615

Camplejohn RS, Macartney JC and Morris RW (1989) Measurement of S-phase fractions in lymphoid tissue comparing fresh versus paraffin-embedded tissue and 4′,6′-diamidino-2-phenylindole dihydrochloride versus propidium iodide staining. Cytometry 10:410–416

Clark GM, Dressler LG, Owens MA, Pounds G, Oldaker T and McGuire WL (1989) Prediction of relapse or survival in patients with node-negative breast cancer by DNA flow cytometry. N Engl J Med 320:627–633

Coon JS, Deitch AD and de Vere White RW, et al. (1988) Interinstitutional variability in DNA flow cytometric analysis of tumors. Cancer 61:126–130

Costa A, Mazzini G, Del Bino G and Silvestrini R (1981) DNA content and kinetic characteristics of non-Hodgkin's lymphoma: determined by flow cytometry and autoradiography. Cytometry 2:185–188

Crissman HA and Steinkamp JA (1990) Cytochemical techniques for multivariate analysis of DNA and other cellular constituents. In: Melamed MR, Lindmo T, Mendelsohn ML (eds) Flow cytometry and sorting, 2nd edn. Wiley–Liss, New York

Crissman JD, Zarbo RJ, Ma CK and Visscher DW (1989) Histopathologic parameters and DNA analysis in colorectal adenocarcinomas. Pathology Annual 24(2):103–147

Darzynkiewicz Z (1990) Probing nuclear chromatin by flow cytometry. In: Melamed MR, Lindmo, T, Mendelsohn ML (eds) Flow cytometry and sorting, 2nd edn. Wiley–Liss, New York

Darzynkiewicz Z and Kapuscinski J (1990) Acridine orange: a versatile probe of nucleic acids and other cell constituents. In: Melamed MR, Lindmo T, Mendelsohn ML (eds) Flow cytometry and sorting, 2nd edn. Wiley–Liss, New York

Dean PN (1990) Data processing. In: Melamed MR, Lindmo T, Mendelsohn ML (eds) Flow cytometry and sorting, 2nd edn. Wiley–Liss, New York

Drach J, Gattringer C, Glassl H, Schwarting R, Stein H and Huber H (1989) Simultaneous flow cytometric analysis of surface markers and nuclear Ki-67 antigen in leukemia and lymphoma. Cytometry 10:743–749

Ensley JF, Maciorowski Z, Pietraszkiewicz H et al. (1987) Solid tumor preparation for flow cytometry using a standard murine model. Cytometry 8:479–487

Friedlander ML, Hedley DW and Taylor IW (1984) Clinical and biological significance of aneuploidy in human tumours. J Clin Pathol 37:961–974

Gillett CE, Camplejohn RS and O'Reilly SM (1990) Specimen preparation and proliferation markers in human breast cancer. J Pathol 160:173A

Hedley DW (1989) Flow cytometry using paraffin-embedded tissue: five years on. Cytometry 10:229–241

Hedley DW, Friedlander ML, Taylor IW, Rugg CA and Musgrove EA (1983) Method for analysis of cellular DNA content of paraffin-embedded pathological material using flow cytometry. J Histochem Cytochem 31:1333–1335

Hemming JD, Quirke P, Womack C, Wells M, Elston CW and Bird CC (1987) Diagnosis of molar pregnancy and persistent trophoblastic disease by flow cytometry. J Clin Pathol 40:615–620

Hermansen DK, Badalament RA, Fair WR, Kimmel M, Whitmore WF and Melamed MR (1989) Detection of bladder carcinoma in females by flow cytometry and cytology. Cytometry 10:739–742

Hiddemann W, Schumann J and Andreeff M, et al. (1984) Convention on nomenclature for DNA cytometry. Cytometry 5:445–446

Kallioniemi O-P (1988) Comparison of fresh and paraffin-embedded tissue as starting material for DNA flow cytometry and evaluation of intratumor heterogeneity. Cytometry 9:164–169

Kallioniemi O-P, Blanco G, Alavaikko M et al. (1988) Improving the prognostic value of DNA flow cytometry in breast cancer by combining DNA index and S-phase fraction. Cancer 62: 2183–2190

Koss LG, Wersto RP, Simmons DA, Deitch D, Herz F and Freed SZ (1989) Predictive value of DNA measurements in bladder washings. Cancer 64:916–924

Kurki P, Ogata K and Tan EM (1988) Monoclonal antibodies to proliferating cell nuclear antigen (PCNA)/cyclin as probes for proliferating cells by immunofluorescence microscopy and flow cytometry. J Immunol Methods 109:49–59

Latt SA and Langlois RG (1990) Fluorescent probes of DNA microstructure and DNA synthesis. In: Melamed MR, Lindmo T, Mendelsohn ML (eds) Flow cytometry and sorting, 2nd edn. Wiley–Liss, New York

Lincoln ST and Bauer KD (1989) Limitations in the measurement of c-*myc* oncoprotein and other nuclear antigens by flow cytometry. Cytometry 10:456–462

Linden WA, Köllermann M and König K (1980) Flow cytometric and autoradiographic studies of human kidney carcinomas surgically removed after preirradiation. Br J Cancer 41(suppl IV):177–180

Macartney JC and Camplejohn RS (1990) DNA flow cytometry of non-Hodgkin's lymphomas. Europ J Cancer 26:635–637

Macartney JC, Camplejohn RS, Alder J, Stone MG and Powell G (1986) Prognostic importance of DNA flow cytometry in non-Hodgkin's lymphomas. J Clin Pathol 39:542–546

Macartney JC, Camplejohn RS, Morris R, Hollowood K, Clarke D and Timothy A (1991) DNA flow cytometry of follicular non-Hodgkin's lymphoma. J Clin Pathol 44:215–218

McDivitt RW, Stone KR, Craig RB, Palmer JO, Meyer JS and Bauer WC (1986) A proposed classification of breast cancer based on kinetic information. Cancer 57:269–276

McIntire TL, Goldey SH, Benson NA and Braylan RC (1987) Flow cytometric analysis of DNA in cells obtained from deparaffinized formalin-fixed lymphoid tissues. Cytometry 8:474–478

Melamed MR and Staiano-Coico L (1990) Flow cytometry in clinical cytology. In: Melamed MR, Lindmo T, Mendelsohn ML (eds) Flow cytometry and sorting, 2nd edn. Wiley–Liss, New York

Melamed MR, Mullaney PF and Shapiro HM (1990a) An historical review of the development of flow cytometers and sorters. In: Melamed MR, Lindmo T, Mendelsohn (eds) Flow cytometry and sorting, 2nd edn. Wiley–Liss, New York

Melamed MR, Lindmo T and Mendelsohn ML (eds) (1990b) Flow cytometry and sorting, 2nd edn. Wiley–Liss, New York

Merkel DE and McGuire WL (1990) Ploidy, proliferative activity and prognosis. DNA flow cytometry of solid tumors. Cancer 65:1194–1205

Muss HB, Kute TE and Case LD, et al. (1989) The relation of flow cytometry to clinical and biologic characteristics in women with node negative primary breast cancer. Cancer 64:1894–1900

O'Reilly SM, Camplejohn RS and Barnes DM, et al. (1990a) DNA index, S-phase fraction, histological grade and prognosis in breast cancer. Br J Cancer 61:671–674

O'Reilly SM, Camplejohn RS, Barnes DM, Millis RR, Rubens RD and Richards MA (1990b) Node negative breast cancer: prognostic subgroups defined by tumour size and flow cytometry. J Clin Oncol 8:2040–2046

Ormerod MG (ed) (1990) Flow cytometry. A practical approach. IRL Press, Oxford

Ormerod MG, Imrie PR (1991) The use of multiparametric analysis when recording a DNA histogram using flow cytometry. Cytometry (in press)

Pallavicini MG, Taylor IW and Vindelov LL (1990) Preparation of cell/nuclei suspensions from solid tumors for flow cytometry. In: Melamed MR, Lindmo T, Mendelsohn ML (eds) Flow cytometry and sorting, 2nd edn. Wiley–Liss, New York

Price J and Herman CJ (1990) Reproducibility of FCM DNA content from replicate paraffin block samples. Cytometry 11:845–847

Raber MN and Barlogie B (1990) DNA flow cytometry of human solid tumors. In: Melamed MR, Lindmo T, Mendelsohn ML (eds) Flow cytometry and sorting, 2nd edn. Wiley–Liss, New York

Rehn S, Glimelius B, Strang P, Sundström C and Tribukait B (1990) Prognostic significance of flow cytometry studies in B-cell non-Hodgkin's lymphoma. Hematol Oncol 8:1–12

Remvikos Y, Laurent-Puig P, Salmon RJ, Frelat G, Dutrillaux B and Thomas G (1990) Simultaneous monitoring of p53 protein and DNA content of colorectal adenocarcinomas by flow cytometry. Int J Cancer 45:450–456

Schutte B, Reynders MMJ, Bosmann FT and Blijham GH (1985) Flow cytometric determination of DNA ploidy level in nuclei isolated from paraffin-embedded tissue. Cytometry 6:26–30

Silvestrini R, Costa A, Veneroni S, Del Bino G and Persici P (1988) Comparative analysis of different approaches to investigate cell kinetics. Cell Tissue Kinet 21:123–131

Steen HB (1990) Characteristics of flow cytometers. In: Melamed MR, Lindmo T, Mendelsohn ML (eds) Flow cytometry and sorting, 2nd edn. Wiley–Liss, New York

Stephenson RA, Gay H, Fair WR and Melamed MR (1986) Effect of section thickness on quality of flow cytometric DNA content determinations in paraffin-embedded tissues. Cytometry 7:41–44

Toikkanen S, Joensuu H and Klemi P (1989) The prognostic significance of nuclear DNA content in invasive breast cancer – a study with long-term follow-up. Br J Cancer 60:693–700

Toikkanen S, Joensuu H and Klemi P (1990) Nuclear DNA content as a prognostic factor in $T_{1-2}N_0$ breast cancer. Am J Clin Pathol 93:471–479

Uyterlinde AM, Baak JPA, Schipper NW, Peterse H, Matze E and Meijer CJL (1990) Further evaluation of the prognostic value of morphometric and flow cytometric parameters in breast cancer patients with long follow-up. Int J Cancer 45:1–7

Van Dilla MA, Trujillo TT, Mullaney PF and Coulter JR (1969) Cell microfluorometry: a method for rapid fluorescence measurement. Science 163:1213–1214

Vindelov LL and Christensen IJ (1990) A review of techniques and results obtained in one laboratory by an integrated system of methods designed for routine clinical flow cytometric DNA analysis. Cytometry 11:753–770

Waggoner AS (1990) Fluorescent probes for cytometry. In: Melamed MR, Lindmo T, Mendelsohn ML (eds) Flow cytometry and sorting, 2nd edn. Wiley–Liss, New York

Watson JV, Sikora K and Evan GI (1985) A simultaneous flow cytometric assay for c-*myc* oncoprotein and DNA in nuclei from paraffin embedded material. J Immunol Methods 83:179–192

Watson JV, Stewart J, Evan GI, Ritson A and Sikora K (1986) The clinical significance of flow cytometric c-*myc* oncoprotein quantitation in testicular cancer. Br J Cancer 53:331–337

7 Measurement of Cell Proliferation Using Bromodeoxyuridine

G.D. WILSON and N.J. McNALLY

This chapter is dedicated to the memory of Nic
McNally, who died shortly after its completion. As well
as being a skilled scientist, Nic was a charming and
gracious man. I owe everything to Nic for my own
scientific development. He was respected by all and will
be sadly missed by his friends and colleagues.

G.D.W.

Introduction

Cellular proliferation is an integral part of the maintenance of the organism. While many tissues in the adult are essentially non-dividing, others, the so-called cell renewal systems, are in a constant state of division, maturation and loss, for example in skin and gut. The process of homeostasis keeps the balance between cell production and cell loss so that the cell proliferation rate is exactly balanced by cell loss. It is a fundamental part of the development of cancer that there is a loss of this homeostatic control such that cell production becomes uncontrolled, with cells proliferating faster than they are lost. It is now generally recognised that the rate of tumour cell proliferation can be a major factor in determining the success of treatment. This is certainly the case for the treatment of tumours by fractionated radiotherapy (Fowler 1986; Withers et al. 1988) and is expected to be true for chemotherapy when there are also intervals between treatments. In the case of radiotherapy, clinical trials are underway to assess strategies to overcome tumour cell proliferation during treatment by means of accelerated fractionation schedules when, in the extreme, the overall treatment time has been reduced to 12 days without a break, giving three fractions per day (Saunders et al. 1988). Of course, cell proliferation during treatment will not be a problem with all tumours. Many may be satisfactorily treated by conventional schedules. It would clearly be of considerable value, therefore, if those tumours for which cell proliferation may be a problem could be identified prior to treatment, allowing a rational selection of patients for accelerated treatment (Wilson et al. 1988).

These considerations emphasise the concern in this chapter with measuring temporal parameters, rates of cell production, and duration of specific cell cycle phases. Other chapters in this book deal with a number of cell kinetic parameters which are indirect indicators of rates of cell division, e.g. MI, LI, SPF and growth fraction markers. Here, we shall describe techniques using BUdR labelling of DNA synthesising cells, combined with flow cytometry and immunohistochemistry, to measure parameters of cell proliferation, and in particular the LI, duration of S phase (T_S) and hence the potential doubling time (T_{pot}). The emphasis is on the application to human tumours, using in vivo labelling, to obtain information on cell proliferation that will be of predictive value, to aid the clinician in designing the most appropriate form of treatment for the individual patient.

Autoradiography using radioactive precursors of DNA has traditionally been the mainstay of techniques to measure cell kinetics and proliferation in tumours and normal tissues. This has been of great value in experimental systems, although the procedures are time-consuming and tedious and can be difficult to apply in very slowly proliferating tissues. However, autoradiographic techniques are clearly not generally applicable to human tissues and tumours except in limited studies, for example of labelling index assessed by in vitro labelling (for reviews see Meyer 1982; Tubiana and Courdi 1989). The BUdR technique, however, involves a non-radioactive label which can be injected into humans. Subsequent biopsies can be analysed rapidly by flow cytometry for their DNA and BUdR content and, as we shall show, the proliferation parameters outlined above can be derived. In the case of aneuploid tumours, the aneuploid component can be analysed separately from the diploid cells. Immunohistochemical techniques can also be used to examine the localisation of BUdR-labelled cells in tissue sections both of tumours and, in surgically resected samples for instance, adjacent normal tissue. This can provide information on the structural distribution of proliferating cells in relation to differentiation status, vascular components, etc., as well as allowing assessment of microscopic heterogeneity.

These techniques were made possible by the development of monoclonal antibodies recognising the halogenated pyrimidines incorporated into DNA (Gratzner 1982) and by FCM methods to measure simultaneously the uptake of BUdR and total DNA content (Dolbeare et al. 1983). The use of a monoclonal antibody for detection meant that the DNA precursor (BUdR) did not have to be radiolabelled, as was the case for ^{3}H-thymidine and autoradiography. This created the possibility of in vivo usage to exploit techniques developed at the Gray Laboratory to measure both LI and T_S, and thus T_{pot}, from a single observation made several hours after the administration of BUdR (Begg et al. 1985). The administration of BUdR to patients and the use of FCM for detection satisfies many of the requirements for a predictive measurement, and combined with immunohistochemistry, has opened new opportunities to study cell proliferation in vivo in man as well as in experimental systems.

The Parameters of Tumour Cell Proliferation

As we have indicated, the parameters of interest in this chapter are those which best describe the proliferative characteristics of a tumour, since we wish to identify those measurements which predict the ability to undergo rapid (compensatory) proliferation during treatment. Many clinical studies have demonstrated that such proliferation does take place, having serious clinical consequences (Taylor et al. 1990; Withers et al. 1988). Fowler (1986) and others have identified T_{pot} as the best predictive parameter.

Tumours consist of both dividing and non-dividing cells, and T_{pot} is the time for such a population to double its number (Steel 1977). In untreated tumours, the equivalent parameter is the volume doubling time which is very much longer than T_{pot} because of the existence of cell loss from the tumour. However, clinical studies have shown that during fractionated radiotherapy the doubling time is reduced, in many cases to values approaching the potential doubling time (Fowler 1986).

The potential doubling time is given by the formula:

$$T_{pot} = \frac{\lambda T_S}{LI}. \tag{1}$$

where λ is a correction factor for the age distribution of the cell population and typically varies between 0.693 and 1.0. A value of 0.8 has been chosen for our studies, based on calculations in experimental tumours. A full discussion of this formula and procedures to derive T_{pot} can be found below.

Flow Cytometric Methods

One of the advantages of the BUdR–FCM is that the time-scale of obtaining a result can be in the region of 1–2 days following biopsy. Large numbers of cells, typically 10000, are analysed in a quantitative manner. In terms of cost, a flow cytometer is required, although this does not have to be on-site as the tissue preparation procedures mean that samples can be sent by post to a centre with a FCM. Clinically, BUdR has proven non-toxic in the doses administered (200 mg) for cell kinetic studies. The fact that the information can be obtained from one biopsy means that patients do not have to undergo further surgical procedures other than those required for routine histopathologic diagnosis. The only inconvenience is a single bolus injection of 200 mg of BUdR given 4–8 h prior to surgery. The half-life of BUdR is short (c. 10–15 min) such that it will only be incorporated into S-phase cells at the time of injection analogous to pulse-chasing in vitro.

Tissue Preparation and Staining

The tissue preparation and staining methods to identify BUdR-labelled cells in human tumours are outlined in Table 7.1. There are several important aspects to the procedure. Firstly, the use of pepsin to produce nuclei from solid pieces of tumour has meant that fixed material can be used with analysis at one centre. It has also meant that the yield of nuclei per gram of tumour has greatly increased in comparison to mechanical or enzymatic methods to produce single

Table 7.1. Procedure for preparation of nuclei from solid tumours and staining for BUdR incorporation and total DNA content

---

1. Fix biopsy or resection material in 70% ethanol as solid pieces no bigger than 0.5 cm; store at 4°C until staining
2. Remove from alcohol and mince tissue (up to 200 mg) with scissors into 1 mm pieces
3. Add 8 ml of 0.4 mg/ml pepsin (Sigma) in 0.1 M HCl (the pepsin should first be dissolved in a small amount of phosphate-buffered saline – PBS) to the tissue fragments
4. Incubate for up to 60 min at 37°C with constant agitation: the incubation time will depend upon the individual tumour; typically, 30 min is sufficient for dissociation into nuclei
5. Filter nuclei suspension through 35 μm nylon mesh
6. Centrifuge at 2000 r.p.m. for 5 min and discard supernatant
7. Resuspend pellet in 2.5 ml 2 M HCl for 12 min at room temperature to partially denature DNA to allow access of the monoclonal antibody to its binding sites
8. Add 5 ml PBS and centrifuge at 2000 r.p.m. for 5 min
9. Discard supernatant and resuspend in 5 ml PBS; spin at 2000 r.p.m. for 5 min and discard supernatant
10. Resuspend pellet in 500 μl of PBS containing 0.5% normal goat serum and 0.5% Tween-20; add 25 μl of anti-rat BUdR monoclonal antibody supernatant (Sera Lab) and incubate for 1 h at room temperature with occasional mixing
11. Add 5 ml PBS, spin at 2000 r.p.m. and discard supernatant
12. Resuspend pellet in 0.5 ml PBS–Tween–normal goat serum (NGS) containing 20 μl of goat anti-rat 1 gG FITC conjugate; incubate for 30 min at room temperature
13. Add 5 ml PBS, spin 2000 at r.p.m. for 5 min and discard supernatant
14. Resuspend pellet in 2 ml PBS containing 10 μg/ml PI
15. Analyse on flow cytometer

---

cells. This has led to the measurement being feasible on drill biopsy material (20–100 mg). It also means that the dissociation procedure is the same for each tumour type. The quality of both the BUdR and DNA staining has been improved, as the removal of cytoplasm abolishes the majority of non-specific staining of the antibodies and improves the CVs of the DNA distributions.

The critical step in the procedure is the requirement to partially denature DNA with HCl to allow access of the monoclonal antibody to its binding sites. This step must achieve sufficient unwinding of DNA to be able to detect BUdR incorporation, but must not unwind it to such an extent that the stoichiometry of PI intercalation into the DNA is compromised. Through experience, we have arrived at a denaturation time of 12 min. This may be optimal for some tumours but not for others because there is variation in the denaturability of different tumours and normal tissues. However, it is not possible to optimise the denaturation period for each tumour as there is often only enough material for one staining. The starting number of nuclei should not exceed 4×10^6 otherwise more monoclonal antibody should be used. If the detection of BUdR is low, it is always possible to repeat steps 9–14 to try to enhance the fluorescence signal. The major disadvantage of using nuclei is that the opportunities for simultaneous staining of BUdR with another marker, such as a cytokeratin, are restricted to nuclear proteins that can withstand the staining procedure.

Data Acquisition

We collect the data on an Ortho Cytofluorograph Systems 50-H or a Becton–Dickinson FACScan, but the majority of commercial flow cytometers are capable of acquisition and analysis of BUdR–DNA profiles. Routinely, laser excitation is achieved using 200 mW of 488 nm (only 15 mW required on a FACScan). In the Ortho FCM, fluorescence is collected orthogonal to the incident beam using a microscope objective with a low numerical aperture. The exciting wavelength is deflected using a 510 nm long-pass dichroic filter. Green fluorescence is collected using a 530 nm bandpass filter (510–560 nm) and red fluorescence deflected using a 590 nm short-pass dichroic and a 620 nm long-pass filter. In the FACScan, BUdR-related fluorescence (FITC) is collected into FL1 and the DNA signal (PI) is collected into FL2. This latter channel is designed for phycoerythrin, which means that compensation is required to subtract spectral overlap of FL1 from FL2, typically 5%–15% depending on the intensity of the FITC signal. It is important, when analysing DNA, to be capable of discriminating cell doublets that may interfere with the DNA profile; this is particularly important when dealing with tumours and tissues having a low LI. Most commercial machines offer the option of pulse processing, in which the DNA signal can be simultaneously analysed as an area (integral), peak (height or amplitude) and width signal. Combinations of these parameters can separate two G_1 cells, passing coincidentally through the laser beam, from a G_2 cell.

The number of nuclei collected is usually 10 000, but this should be increased if the LI of the population is low. It is advisable to collect data at no more than 100 events per second to achieve the best CVs on the DNA profile. Routinely, we collect the FITC signal on a linear scale, but this can be changed to logarithmic if there is a wide variation in green fluorescence. Data are usually acquired in list mode for subsequent analysis.

Data Analysis

Figure 7.1a shows a typical profile of a diploid human tumour labelled with BUdR in vivo and 7.1b shows an aneuploid tumour. The diploid tumour was a follicular lymphoma removed 5.7 h after the injection of BUdR. Interpretation of the bivariate distribution of BUdR versus DNA is straightforward. It is evident that BUdR-labelled cells have redistributed through the cell cycle. Some have divided in the 5.7 h period, and reside in G_1/G_0 at channel 30. These are distinguishable from other G_1/G_0 cells (that have not yet progressed into S phase) by virtue of their BUdR incorporation. The other BUdR-labelled cells still reside in S and G_2+M, but they have progressed through S phase since the time of labelling when there would have been a uniform distribution of BUdR-labelled cells throughout S phase between the G_1 and G_2 DNA contents. It is this redistribution of BUdR-labelled cells in the time between injection and biopsy that forms the basis of the analytical procedures that have been developed to estimate T_S, LI and T_{pot}.

The Relative Movement Method

The original procedure to calculate T_{pot} from a single biopsy was devised by Begg et al. (1985). The method termed "relative movement" (RM) depends on several critical assumptions. The first of these is that the initial distribu-

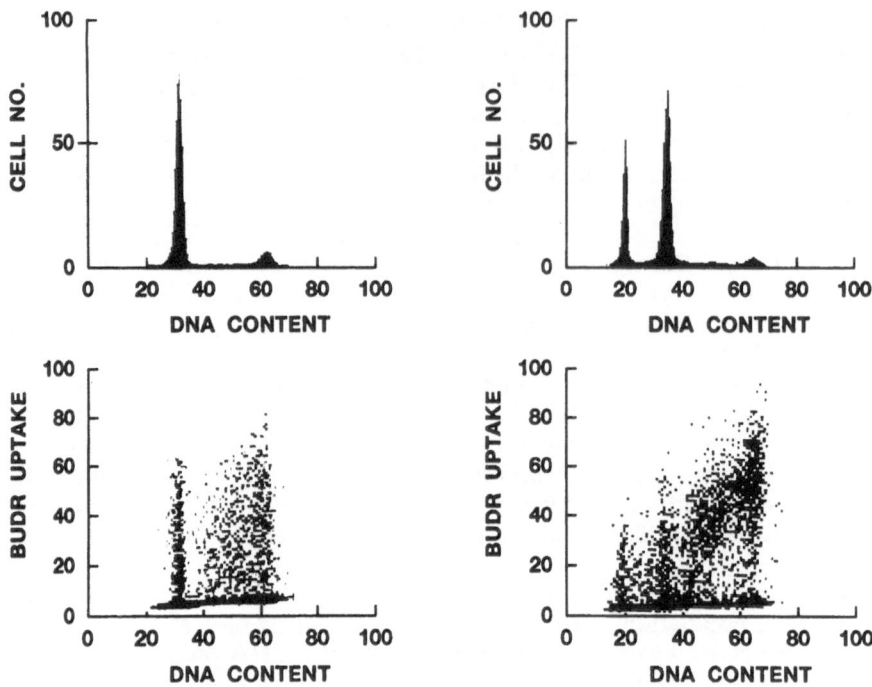

Figure 7.1. DNA profiles of a diploid (*left*) and an aneuploid (*right*) human tumour. The diploid tumour was a follicular lymphoma removed 5.7 h after injection of BUdR. The aneuploid tumour was a malignant melanoma biopsied 6.2 h after injection. *Lower*: the bivariate distributions of BUdR uptake versus total DNA content for these two tumours.

tion of BUdR-labelled cells obtained immediately after injection is uniform throughout the S phase (see Fig. 7.2). If this is the case, a region set around all the BUdR-labelled cells would give a mean on the x-axis (DNA content), in mid-S, halfway between the G_1 and G_2 DNA contents. The term relative movement comes from expressing the mean DNA content of the labelled S-phase cells relative to the difference in DNA content between the G_1 and G_2 cell. This is simply done by subtracting the mean DNA content of the G_1 population from that obtained for the BUdR-labelled cells and dividing it by the mean DNA content of the G_1 subtracted from the G_2 cells:

$$RM = \frac{FL(BUdR) - FL_{G_1}}{FL_{G_2+M} - FL_{G_1}}. \tag{2}$$

The initial RM value would be 0.5 under the criteria described above.

The second assumption states that the progression of BUdR-labelled cells through the S phase is linear with time. The RM calculation is made only on the cohort of BUdR-labelled cells which continue to progress through S and G_2+M and not those which divide and enter G_1. This results in an increase in RM from 0.5 with time. The theoretical progress of BUdR-labelled cells is shown in Fig. 7.2. It is evident that the RM term will equal 1.0 when the BUdR-labelled cells that were at the beginning of S at $t = 0$ have reached G_2+M but not divided, i.e. in a time equal to T_S. Thus, assuming linearity, a

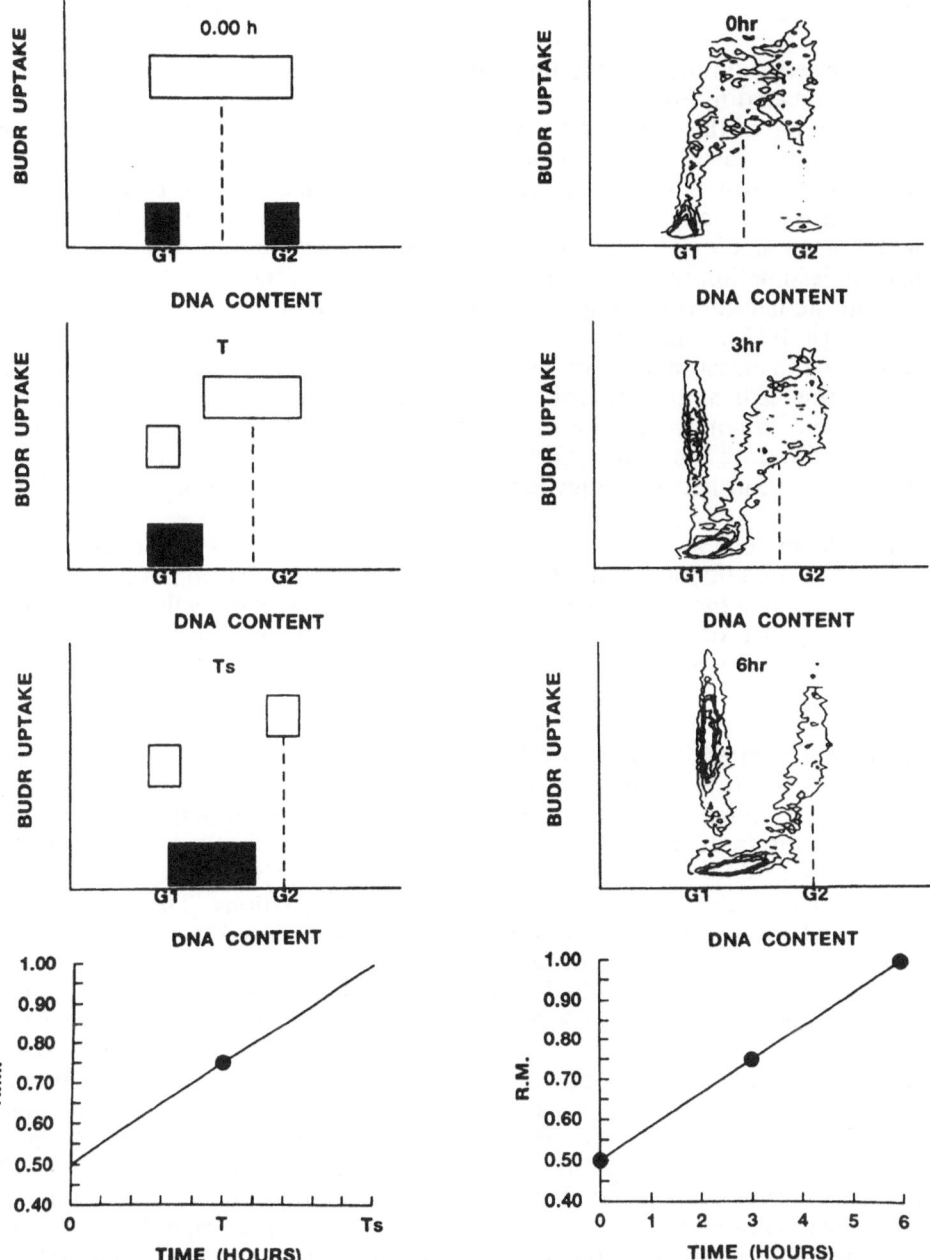

Figure 7.2. Calculation of the duration of S phase, labelling index and potential doubling time from a single biopsy. *Left*: the theoretical distribution of BUdR pulse-labelled cells (plain boxes) as they distribute around the cell cycle. At time 0 there is a uniform distribution between G_1 and G_2. At a time T, several hours later, the BUdR-labelled cohort increases its DNA content and shifts towards G_2. In addition, some BUdR-labelled cells divide and enter G_1. At time equal to T_S, the BUdR-labelled cells now reside only in G_2 or in G_1. The dashed line represents the mean value for DNA content of the BUdR-labelled cohort which has not divided and entered G_1. The graph represents the extrapolation of T_S using the observation made at time T and the assumed starting value of 0.5. *Right*: shows comparable actual data obtained from chinese hamster V79 cells pulse-labelled with 10 µM BUdR for 20 min and samples at time intervals thereafter. A value for T_S, of 6 h, was obtained from these data from the relative movement plot.

single observation made some time after the injection allows the calculation of T_S, as indicated in Fig. 7.2.

The timing of this observation is important. At shorter times, there may not be sufficient redistribution of cells through S phase to be able to discriminate those cells which have divided from those which are moving out of early S phase. If no division of BUdR-labelled cells is observed, then the estimation of RM is not reliable. This may indicate a long G_2+M period which will tend to bias the estimation of T_S to a short value due to persistence of BUdR-labelled cells in the late-S/G_2+M population, producing a RM value weighted towards 1.0 (Begg 1989). Similarly, too long a period between injection and biopsy results in the measurements of RM of the BUdR-labelled cells being performed on too few cells as the majority will have divided and not be included in the calculation. When profiles are obtained in which the BUdR-labelled cells have all moved out of S to either G_2+M or G_1, then the estimate of T_S will equal the delay period between injection and biopsy and can only be an upper limit for T_S.

In profiles such as that in Fig. 7.1a the estimation of RM is uncomplicated. A region would be set around the BUdR-labelled cohort which has not divided, extending from the right-hand limit of the BUdR-labelled cells in G_1 to the right-hand edge of the G_2+M cells. This region extends to the upper limit of the display screen and its lower limit is defined by the boundary between those cells which show significant BUdR uptake from those which do not. This can be assessed either by staining the cells not exposed to BUdR or by omitting the monoclonal antibody from cells that have incorporated BUdR to obtain a "negative" control profile. However, it is common in solid tumour biopsies not to have enough material to stain twice. The lower limit delineating BUdR-labelled cells from non-labelled cells has to be set subjectively. This seldom presents a problem in extracted nuclei, due to the abolition of non-specific staining and a clear distinction between these populations. The mean DNA content of the G_1 and G_2+M cells can either be deduced from the bivariate profile or the single parameter DNA profile.

In both diploid and aneuploid tumours, it is our convention always to set the region adjacent to the right-hand edge of G_1, if possible, to determine the relative movement of the BUdR-labelled cells. However, in some aneuploid tumours, this is not always the case. Fig. 7.1b shows the bivariate distribution obtained from a melanoma removed 6.2 h after BUdR injection. In this tumour, the DNA index was 1.8. Moreover, there was significant proliferation in both the diploid and aneuploid subpopulations, resulting in overlapping BUdR profiles. This would suggest that both populations represent different tumour stem lines. However, it also means that a region set to the right-hand edge of the aneuploid G_1 population, to calculate the RM of the aneuploid BUdR-labelled cells, would include some diploid S and G_2+M cells. In circumstances such as these, a subjective decision has to be made as to whether this region can be shifted to the right to exclude the contaminating diploid cells. In cases such as Fig. 7.1b this is clearly feasible, as a cohort of aneuploid BUdR-labelled cells still progressing through S phase has shifted their DNA content sufficiently towards G_2+M to be easily discriminated from the diploid G_2+M cells. In instances such as this, the region would be set to the right-hand edge of the diploid G_2+M. If the discrimination is not clear, then the region should be set as described previously. This may weight the value to a longer T_S if there are overlapping proliferating cells, but it is less subjective than setting a region further to the right and biasing the T_S estimate to a shorter value. This problem only arises in aneuploid tumours which exhibit both diploid and aneuploid proliferation; it is often the case that the vast majority of proliferating cells are found in the aneuploid subpopulation, with little evidence of proliferation in the diploid cells. Again, in an aneuploid tumour, the mean DNA content of the G_1/G_0 cells and the G_2+M cells can be calculated either from the bivariate or the single parameter DNA distribution.

The duration of S phase is derived by:

$$T_S = \frac{1.0 - 0.5}{RM - 0.5} \times t \tag{3}$$

where t is the time between injection and biopsy.

The other parameter required for calculation of T_{pot} is the LI, the fraction of the population synthesising DNA, identified here by their uptake of BUdR. It is evident from the presence of BUdR-labelled cells in G_1 in both profiles in Fig. 7.1 that cells have divided and shared their BUdR between daughter cells. As a result, the number of labelled cells will have expanded compared to those present at the time of injection. This has to be corrected. In the time between injection and biopsy, some BUdR-labelled cells will divide, as will the cells that were originally in G_2+M at the time of labelling. It is difficult to estimate accurately how many G_2+M cells divide since, having divided, they are "lost" in the G_1 population. It is also our experience that a significant proportion (c. 50%) of these cells often remain in G_2+M, even though BUdR-labelled cells do divide. Estimation of G_2+M can be made from the DNA profile, but this depends on good quality DNA staining and computer models for DNA analysis. As a consequence, we do not correct for divided G_2+M cells. In practice, the number is small and would make little difference to the estimate of LI. The division of BUdR-labelled cells is a more significant error and this can be corrected more easily by setting a region around the BUdR-labelled G_1 population. To correct the LI, this number is halved and subtracted from the total number of BUdR-labelled cells and the total cell number. These values are used to calculate the approximate LI at the time of injection.

It is apparent from Fig. 7.1b that in aneuploid tumours the LI can be calculated for the aneuploid subpopulation, in addition to the total population, by setting windows according to DNA content. It is our experience that in such tumours the aneuploid LI is always greater than the total cell LI due to the presence of non-proliferating diploid cells. It is, of course, not possible to discriminate normal from tumour cells in diploid tumours.

Having estimated T_S and LI, the final parameter, the T_{pot}, can be calculated using equation (1). One final assumption is made and that is ascribing a value of 0.8 for λ. This is a complex term describing the age distribution of the cell population (see Chap. 4 and Steel 1977) and is concerned with the position of S phase in the cell cycle. For rapidly dividing populations, the term can be unity or up to 1.38 in extreme situations; the factor approaches 0.693 in steady-state conditions. The value of 0.8 was not arbitrarily chosen; it was adopted from cell cycle analysis of a variety of experimental tumours in which λ could be calculated. In addition, the factor is a constant and so would not affect the ranking of a tumour as rapidly or slowly proliferating; it would only affect the absolute estimate of T_{pot}.

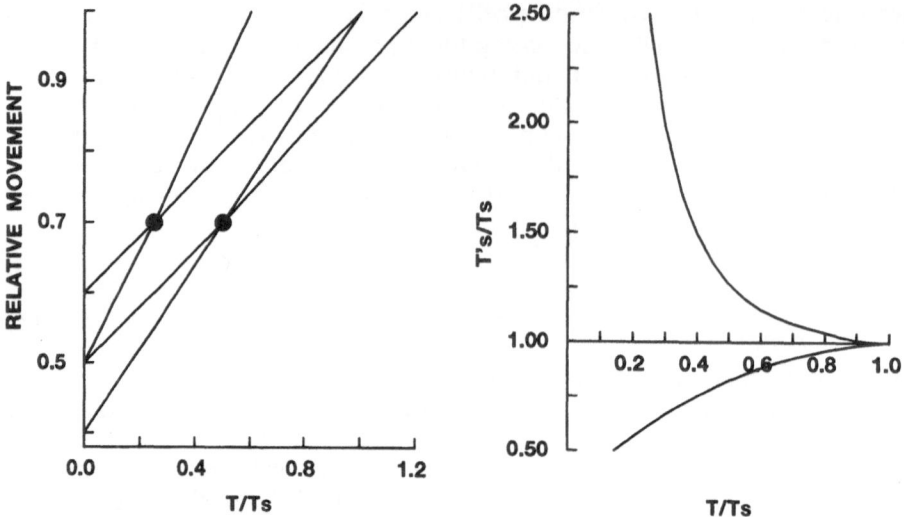

Figure 7.3. Computation of errors in the estimation of T_S from a single observation. *Left*: a schematic plot of relative movement versus time of sampling as a function of T_S using initial intercepts of 0.6, 0.5 and 0.4. The filled symbols represent actual RM observations. Lines have been drawn to show the errors in T_S estimation if an assumed RM(0) value of 0.5 was chosen instead of the situation where the true RM(0) was 0.6 or 0.4. *Right*: the fractional errors (Ts/T_S) using a data set generated, from the left panel, for different RM observations made at increasing intervals of T/T_S in which an assumed RM(0) of 0.5 was used when the actual RM(0) was 0.4 (upper line) or 0.6 (lower line). The closer the observation is made to 0.5 T_S or greater, the smaller the errors.

The two principal assumptions in the RM calculation are that the movement of labelled cells through S is linear with time and that the initial RM value is 0.5. The assumption of linearity (beyond T_{G_2+M}) is not strictly correct, and modifications of this approach are discussed below. The error in assuming an initial RM value of 0.5 when it might be (say) 0.4 or 0.6 is maximum at early sampling times and becomes progressively less as the sample time approaches T_S. This is illustrated in Fig. 7.3 (left). Looking at the plots for initial values of 0.6 and assuming 0.5, it is clear that as the sampling time gets larger, the difference between the extrapolation to an RM of 1.0 gets smaller and smaller so that, irrespective of the RM(0) value chosen, the error in estimating T_S gets smaller. The same is true for a correct RM(0) of 0.4 but an assumed value of 0.5. Fig. 7.3 (right) shows that in order to be within 25% of the correct T_S when RM(0) may be between 0.6 and 0.4, we need to sample at times greater than 0.4 T_S. Of course, the closer RM(0) is to 0.5, the less critical this timing will be. It is our experience that in experimental tumours, and a number of normal murine tissues, that a value of 0.5 is usually a good estimate of the true value, which is sometimes closer to 0.6 but not less than 0.5.

Of course, if it were possible to take two biopsies then it would be possible to measure RM(0) as well as a value at a later time. A further possibility is to use a second label, such as IUdR, as well as BUdR. Monoclonal antibodies are now available having different binding affinities to these two DNA precursors. By giving one drug at $t = 0$ and another just before biopsy, two values of RM can be derived. However, there may be ethical problems in using two drugs in this way, and the flow cytometric procedures are much more complicated, requiring three-colour fluorescence detection, probably with the use of two lasers.

Modifications of the Single Relative Movement Calculation

In a number of theoretical studies, White and his colleagues from Houston have attempted to put the calculation of T_S on a more rigorous mathematical footing. Their approaches will be outlined briefly here and we shall compare results from a given set of data using these different approaches.

The first modification by White and Meistrich (1986), attempted to derive the relative movement at time zero (RM(0)) rather than assume 0.5. In a complex series of derivations based on expressions for exponentially growing populations, the authors were able to show that:

$$RM(0) = \frac{P_S - Z}{Z(P_S + P_{G_2+M})}, \tag{4}$$

where $\quad Z = \ln\left(\frac{1 + P_S + P_{G_2+M}}{1 + P_{G_2+M}}\right).$

P_S and P_{G_2+M} represent the proportion of cells in S phase and G_2+M. These values could be calculated from the DNA profile. The slope (M) of the relative movement plot was shown to be equal to $1/(2T_S)$. Using this and equation (4), T_S could be calculated as

$$RM(t) = RM(0) + t/2T_S, \tag{5}$$

where RM(t) is the relative movement at the observation time t.

In practice, this procedure can present more problems than it solves when considering human solid tumours. The main problems are in estimating the fractions of cells occupying S and G_2+M by computer analysis of the DNA profile. Often the specimens are aneuploid, which means that the estimate of cell cycle phase distributions becomes dependent on the model chosen. In addition, the quality of the DNA profile is sometimes compromised by the necessity to denature DNA for anti-BUdR monoclonal antibody access; this may render it difficult to analyse, the CVs for the different populations being too large for the computer models.

The above method was superseded by the "depletion function" (White 1989). This function ($D^l(t)$) is a transformation of the fraction of BUdR-labelled undivided cells ($f^{lu}(t)$) into a form that is a linear function of time over much of its range. T_{pot} can be calculated directly from $D^l(t)$ using the equation

$$D^l(t) = \ln[1 + f^{lu}(t)] = \ln[2 - f(G_1)] - \frac{\ln(2)t}{T_{pot}}. \tag{6}$$

The same population of labelled undivided cells as was used in the relative movement analysis is used in this derivation, except that it is the fraction of these cells ($f^{lu}(t)$), rather than their mean fluorescence, which is modelled. For observation times greater than T_{G_2+M}, the depletion function is a linear, decreasing function of time, with a slope of $-\ln (2) \, t/T_{pot}$ and an intercept with the y axis that depends on the fraction of all cells in the population that are in $G_1(f(G_1))$. This fraction has to be calculated from a one-dimensional DNA histogram.

The depletion function has the advantage in that it is useful for calculating T_{pot} for long delay periods between injection and biopsy (up to $T_S + T_{G_2+M}$), but it is very dependent upon the accuracy of the estimation of $f(G_1)$. This again would present problems in some aneuploid tumours and in those tumours in which DNA staining has been compromised by denaturation.

The most recent attempt to improve on the RM method is that described by White et al. (1990). This introduces a quantity, v, which is a function of the fraction of labelled divided cells and the fraction of labelled undivided cells. This value relates T_S to T_{pot} and is only weakly dependent on time following labelling. It is calculated using the same regions for relative movement but, as with the depletion function, using fractions instead of means:

$$v = \ln\left[\frac{1 + f^{lu}(t)}{1 - f^{ld}(t)/2}\right], \qquad (7)$$

where $f^{lu}(t)$ is fraction of cells that are labelled and remain undivided at time t, and $f^{ld}(t)$ are those labelled cells which have divided. White et al. then showed that

$$T_{pot} = \ln(2)\frac{T_S}{v}. \qquad (8)$$

A further application of v is that White showed that it is possible to calculate the initial value of RM making the assumption that T_{G_2+M} is approximately $0.3\,T_S$. Thus for times greater than T_{G_2+M} but shorter than $T_S + T_{G_2+M}$:

$$RM(0) = \left[\frac{1 - e^{-v} - ve^{-1.3v}}{v(1 - e^{-1.3v})}\right]. \qquad (9)$$

The T_S and T_{pot} can be calculated by relative movement analysis using equations (5), (7), (8) and (9).

The advantage of v is that it obviates the need to correct the LI for cell division (as described previously), and no assumptions about the value of the term λ are needed.

Comparison of Analytical Methods in the Context of Solid Tumours

The accuracy of each method of determining T_{pot}, T_S and LI can be tested only in experimental systems where multiple samples can be studied. The situation in human tumours is somewhat different in that usually only one biopsy is possible. Table 7.2 presents the data obtained from six different solid tumours, collected at varying times after injection and calculated according to the simple RM and the White et al. v function. The RM method tends to underestimate T_S when compared to the complete calculation using v and RM(0). The variation in T_S estimate is due solely to the difference in the term RM(0). The v

Table 7.2. Comparison of relative movement method to calculate T_S and T_{pot} with the v function in six head and neck tumours. $T_{pot(1)}$ was calculated using v and the derived value for RM(0). $T_{pot(2)}$ was calculated assuming RM(0) = 0.5 as was the case for the relative movement procedure

| Code | DNA index | T_{biopsy} (h) | Relative movement | | | | v | | | |
			RM	T_S (h)	LI (%)	T_{pot} (days)	v	RM(0)	T_S (h)	$T_{pot(1)}$ (days)
277	1.0	9.25	0.87	12.7	7.0	6.0	0.068	0.61	17.8	7.6
278	1.0	11.1	0.90	13.5	7.1	6.4	0.069	0.60	18.5	7.7
287	1.0	4.8	0.71	11.2	6.8	5.5	0.066	0.60	21.8	9.5
285	1.5	11.0	0.76	21.1	11.0	6.4	0.104	0.60	34.4	9.6
296	1.7	3.3	0.73	7.1	4.4	5.4	0.043	0.62	15.0	15.0
295	1.7	4.9	0.79	8.6	13.2	2.2	0.124	0.60	12.9	3.0

function, making the assumption that T_{G_2+M} is approximately $0.3\,T_S$, results in RM(0) ranging from 0.60 to 0.62. This results in an estimate of T_S which ranges from 1.4 to 2.1 times greater than the RM method, with a mean increase of 1.6.

T_{pot} does not vary as greatly as T_S; the values ranged from 1.2 to 1.9 times greater than the RM method, with a mean variation of 1.5.

It would appear, therefore, that differences in the estimate of T_{pot} depend to a greater extent on the choice of RM(0) than on v, λ or corrected LI. It can be seen from Table 7.2 that v and LI are highly correlated. The assumptions made that T_{G_2+M} approximates to $0.3\,T_S$ are not fully substantiated on review of the literature. In Steel's 1977 book *Growth kinetics of tumours*, the T_{G_2+M}/T_S ratio varied from 0.16 to 0.5 in 75 different experimental animal tumours. Data for human tumours, from the same book, show a mean T_{G_2+M}/T_S ratio of 0.36. However, substitution of a different ratio into equation (9) does not greatly affect the estimate of RM(0). The discordance between an assumed RM(0) of 0.5 and a calculated RM(0) of approximately 0.6 does not arise because the initial distribution of BUdR-labelled cells is not uniform throughout the S phase. It arises because the relative movement plot is clearly not a linear function of time, but consists of two almost linear components with an inflection at $t = T_{G_2+M}$ due to the presence of BUdR-labelled G_2+M cells in the region used to calculate RM. This should result in an initial steep rise up to $t = T_{G_2+M}$, followed by a flatter slope between $t = T_{G_2+M}$ to T_S. It is the intercept of the second slope, where most observations on human tumours will be made, that approximates to 0.6.

It is our experience that T_{G_2+M} does not appear to be a long phase in human tumours. Using the BUdR–DNA staining approach, we see divided labelled cells even when the biopsy time is as short as 3h. This is at odds with data generated from FLM curves of human tumours (see Chap. 4 and Steel 1977 pp 202–203), in which the mean T_{G_2+M} of different tumours was 7h. There can be no doubt that BUdR-labelled cells are traversing G_2+M quickly but, as we mentioned previously, there does appear to be a population of either slowly cycling or quiescent G_2+M cells that do not divide in the time-scale of the observation period. In addition, upon examination of mitotic cells on tissue sections after immunolocalisation of BUdR, it is evident that some are not BUdR-labelled, even though BUdR-labelled cells appear in G_1; theoretically, all mitoses should be labelled if the observation period is greater than T_{G_2+M}.

The uncertainty of events in G_2+M in human tumours compared to cells in vitro has led us to continue to adopt an RM(0) of 0.5. As the results of Table 7.2 show, the difference in T_{pot} calculated using either the simple RM method or White's more sophisticated approach is small and, in practice, the RM method is more straightforward to use, appropriate regions being more easy to identify. We have also continued to use the simple correction to allow for cell division in calculating the LI, rather than the v function. We believe that, in practice, these "computational" differences will be insignificant, especially considering biological sources of variation.

Intra- and Inter-Tumour Heterogeneity

Any method that relies on a measurement from a single biopsy is open to the criticism that the biopsy may not be representative of the tumour as a whole. This can be addressed in resection material by taking multiple samples from different areas of the tumour. We will describe our experience with colorectal cancer, but it should be borne in mind that different tumour sites and classes will show different patterns of heterogeneity. Indeed, it is our experience that head and neck tumours do not appear to be as heterogeneous as colorectal or bronchus tumours. As we discuss later, the microscopic heterogeneity (assessed by immunohistochemistry) seen within a biopsy may reflect the macroscopic heterogeneity of the tumour as a whole. In addition, the biopsy is always taken from what appears to be a viable area of the tumour and it is imperative that the FCM analysis is carried out on a piece similar to that which undergoes histopathological assessment.

We have reported recently (Rew et al. 1991) a series of 100 human colorectal tumours studied with BUdR in vivo. In 58 of the patients, multiple pieces of tumour were analysed. These consisted of between two and six 0.5-cm wedges excised from the peripheral quadrants and central regions. The mean and upper and lower limits of the range obtained for each parameter (DNA index, LI, T_S and T_{pot}) for each specimen are shown in Fig. 7.4. Thirty of the tumours had aneuploid DNA stem lines and there was significant variation in DNA index in 15 of these tumours. Six of the tumours showed presence of both diploid and aneuploid stemlines. As Fig. 7.4 shows, for each tumour the proliferation parameters showed variation. Interestingly, the least variation was seen in T_S: mean maximum and minimum variation around the mean were 20.9% and −20.4%, respectively; the values for the LI were 32.8% and −28.9%. The T_{pot} showed most variation, with corresponding values of 46.7% and −37.3%.

This intratumour heterogeneity appears to be quite large. However, we have found other sites to show much less heterogeneity. Also, the objective of the measurement must be borne in mind. We are attempting to characterise individual tumours in terms of the likelihood that they may undergo rapid repopula-

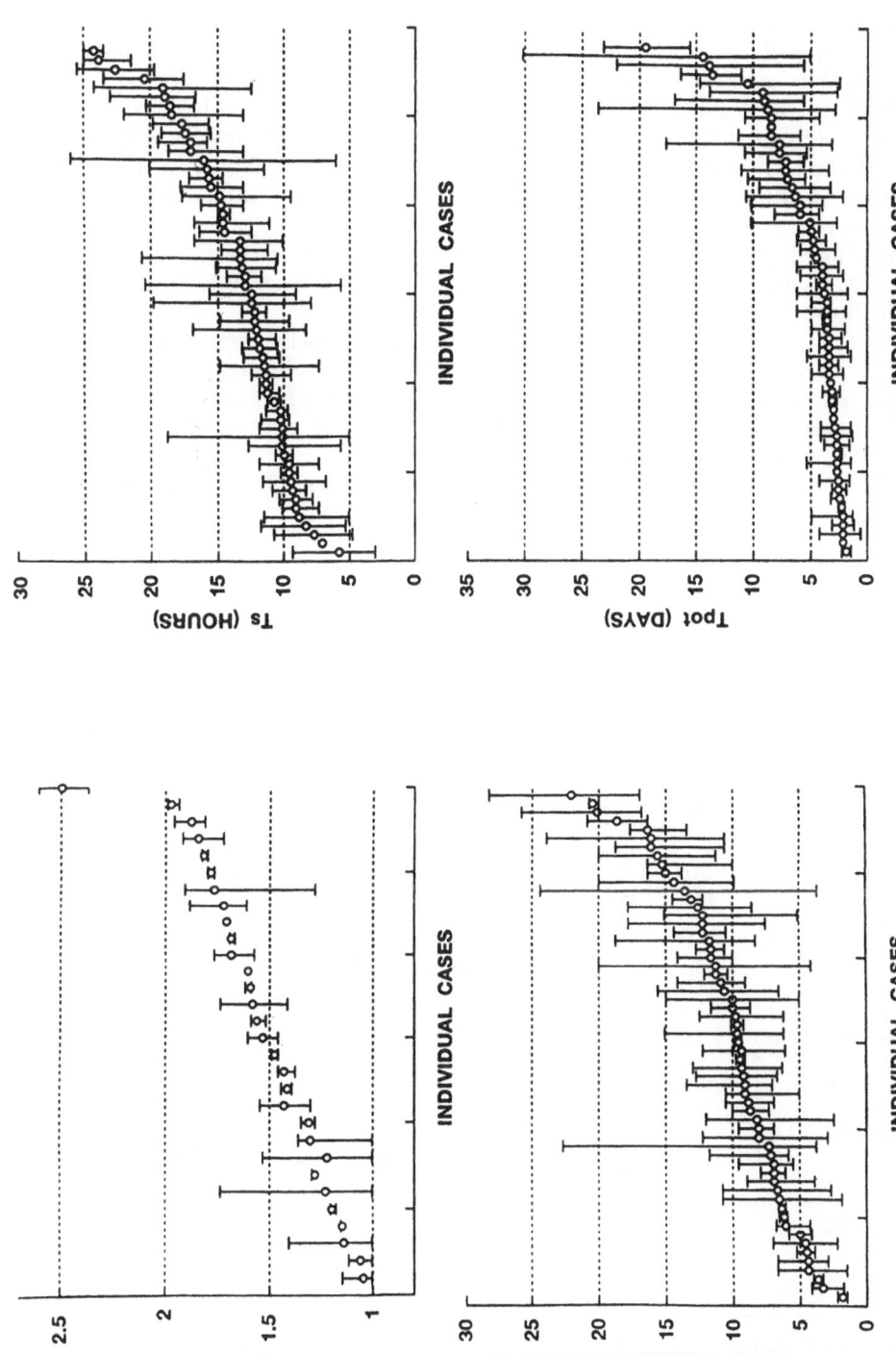

Figure 7.4. Heterogeneity of proliferation parameters in multiple samples from colorectal tumours. The data represent the mean ± range for each individual tumour. The data for DNA indices are quoted for the aneuploid tumours only or those tumours which possessed both diploid and aneuploid stemlines.

tion during treatment. It is likely that prospective studies will show that tumours with a T_{pot} of, say, less than 5 days will have a high risk of undergoing rapid repopulation, those with a value of, say, greater than 9 days won't, and those in between will be more difficult to characterise. Looking at the results in Fig. 7.4, it is apparent that the majority of the tumours would not be compromised in their classification, as described above, by their observed intratumour heterogeneity. Of 38 tumours with a mean T_{pot} of 5 days or less, only 10 contained specimens where the T_{pot} exceeded 5 days. There was a high incidence (10 of 20) of T_{pot} less than 5 days in the tumours with a mean greater than 5 days.

The range is the extreme parameter for the assessment of variation. In many of the cases that transgressed the 5-day boundary, it was only one sample which showed substantial variation.

In many ways, the study described above is an artificial approach to study biopsy variation, as pieces were selected from clearly different parts of the tumour periphery and core. The core specimens may be closer to necrotic areas and less proliferative as a result. An alternative approach is shown in Table 7.3, in which five or six pieces from six tumours were taken from similar "viable" areas of each tumour. The intratumour variation was significantly reduced in all of the proliferation parameters. The coefficient of variation was

Table 7.3. Intratumour variation in cell kinetics for six colorectal tumours. In aneuploid tumours, the aneuploid LI is quoted*

	Sample number						Mean ± SEM	CV (%)
	1	2	3	4	5	6		
Patient 1; DI = 1.0								
LI	5.2	2.4	9.6	4.3	10.7	5.5	6.3 ± 1.3	51
T_S	9.2	7.3	11.3	7.1	10.8	10.0	9.3 ± 0.7	19
T_{pot}	5.9	10.2	3.9	5.5	3.4	6.6	5.9 ± 1.0	41
Patient 2; DI = 1.53								
LI(%)	16.0	9.5	−†	13.2	14.4	10.0	12.6 ± 1.3	22
T_S(h)	13.2	15.6	−	16.7	11.9	15.6	14.6 ± 0.9	14
T_{pot}(d)	2.7	4.3	−	4.2	2.8	3.7	3.5 ± 0.3	21
Patient 3; DI = 1.50								
LI	10.5	12.2	14.7	14.6	14.4	−	13.3 ± 0.8	14
T_S	11.9	13.1	8.0	11.3	11.2	−	11.1 ± 0.8	17
T_{pot}	3.7	3.6	1.8	2.6	2.7	−	2.9 ± 0.3	27
Patient 4; DI = 1.0								
LI	9.6	12.4	13.2	9.5	14.5	−	11.8 ± 1.0	19
T_S	15.7	13.4	16.8	11.6	13.9	−	14.3 ± 0.9	14
T_{pot}	4.3	3.6	4.3	4.0	3.0	−	3.8 ± 0.2	14
Patient 5; DI = 1.41								
LI	9.0	10.8	9.6	12.7	11.0	12.2	10.9 ± 0.6	13
T_S	14.9	17.0	16.3	18.9	18.0	18.1	17.2 ± 0.6	8
T_{pot}	5.4	5.3	5.6	4.9	5.4	5.2	5.3 ± 0.1	4
Patient 6; DI = 1.35								
LI	−	22.9	24.7	18.1	15.7	26.7	21.6 ± 2.1	21
T_S	−	35.1	13.1	17.8	22.7	19.2	21.6 ± 3.7	38
T_{pot}	−	5.2	1.4	3.2	4.7	2.4	3.4 ± 0.7	46

* Units: LI as percentage, T_S in hours, T_{pot} in days.
† −, samples from which it was not possible to make a satisfactory measurement.

close to 20%, although significant variation was seen in patients 1 and 6 in two of the three proliferation parameters. The least variable parameter within a tumour was the T_S.

In practice, for all the sites we have studied, intertumour heterogeneity has been much more significant than intratumour heterogeneity. Our results for six tumour sites are summarised in Table 7.4. The variation between different tumours within the same tumour type is significantly greater than that found within any individual tumour. Of course, it is this intertumour variation (especially in T_{pot}) that lends strength to the expectation that pre-treatment assessment of potential doubling time may be a good predictor of treatment outcome.

Table 7.4 also demonstrates the wide range in T_S in different groups of tumours, which presents difficulties in selecting the appropriate time period between injection of BUdR and surgical biopsy, bearing in mind the considerations in Figure 7.3. The median values of T_S range from around 10 h in head and neck tumours and melanoma to 15 h in cervix and lung. As a general rule, biopsies taken between 4 and 6 h after injection, should yield FCM distributions suitable for analysis with minimal errors. It is possible, with tumours from the cervix and bronchus, to inject late at night and perform surgical procedures early the next day with a delay time of 9 or 10 h. This is not recommended for the other tumour types listed in Table 7.4.

Immunohistochemical Localisation of BUdR

Flow cytometric analysis of BUdR incorporation provides cell kinetic information on human tumours which cannot readily be obtained by other methods. However, it provides "averaged" values and sacrifices information on the tissue spatial distribution of proliferation as a consequence of the prerequisite for a single cell or nuclei suspension. Nevertheless, because BUdR is administered in vivo, the application of immunoperoxidase techniques can be applied to study proliferation at the microscopic level. The information generated by immunohistochemical localisation of incorporated BUdR is essentially identical to that provided by autoradiography. Thus it can not only be a valuable backup should the FCM analysis fail, but also provides information on heterogeneity and structural organisation of proliferation. From a strictly practical point of view, it is our opinion that all biopsies analysed by FCM should be accompanied by an adjacent histological slide, ideally stained for BUdR uptake. This helps to explain unexpected results as well as providing the extra labelling information in the tissue section.

Staining Procedure

Table 7.5 outlines the staining procedure developed by Dr M.H. Bennett and Mrs A. O'Halloran, in the Department of Pathology at Mount Vernon Hospital, for detection of BUdR on human tumour sections. The DNA de-

Table 7.4. Summary of proliferation parameters for six of the major groups of tumours studied at the Gray laboratory (numbers in parentheses are the total number of tumours studied in each group; CVs are percentages)

Tumour group	$LI(\%)$			$T_S(h)$			$T_{pot}(days)$		
	Median	Range	CV	Median	Range	CV	Median	Range	CV
Head/neck (130)	4.9	0.6–20.3	76	9.9	5.4–21.9	45	6.4	1.8–66.6	101
Lung (38)	8.0	0.3–28.5	70	15.1	5.6–37.6	66	7.3	1.4–132.5	167
Oesophagus (30)	7.8	0.4–27.5	64	12.4	6.9–28.6	46	5.2	1.6–56.8	156
Cervix (22)	11.6	2.8–23.4	52	15.8	10.6–30.4	35	4.5	2.9–15.8	61
Melanoma (24)	4.2	1.3–13.6	67	10.7	6.3–20.5	28	7.2	3.5–41.3	82
Colorectal (98)	9.0	0.7–25.5	61	13.1	4.0–28.6	52	3.9	1.7–21.4	68

Table 7.5. Procedure for immunohistochemical localisation of BUdR on human tumour sections

1. Cut 6 μm sections and pick up on poly-1-lysine coated slides and dry for 48 h at 37°C (optional)
2. Take sections through standard deparaffinisation and rehydration with xylene, 100%, 90% and 70% alcohol
3. Block endogenous peroxidase with 0.1% H$_2$O$_2$ in methanol at room temperature for 30 min; wash in tap water, and in distilled water at 37°C
4. Transfer to freshly prepared 0.1% trypsin (BDH) in 0.1% calcium chloride solution (pH 7.8), preheated to 37°C, and incubate for 13 min at 37°C; wash with cold water to stop the action of trypsin
5. Transfer to preheated 1 M HCl at 60°C for 15 min; wash well in tap water
6. Flood slides with 0.01 M Tris buffer with 0.9% saline (TBS) pH 7.6 for 5 min; tip off and wipe around section
7. Add anti-BUdR monoclonal antibody (Dako), diluted 1:30 in TBS containing 1% human AB serum; incubate for 1 h at room temperature; wash 3 times for 1 min with TBS; wipe around sections
8. Add biotinylated rabbit anti-mouse antibody (Dako) diluted 1:300 in TBS containing 1% human AB serum for 1 h at room temperature; wash 3 times in TBS and wipe slide
9. Add avidin biotin complex (ABC) (Vector Labs) for 1 h; wash 3 times in TBS
10. Add diaminobenzidine solution (DAB) (7.5 mg DAB and 15 μl of H$_2$O$_2$ in 100 ml of 0.1 M Tris buffer pH 7.6) for 10 min; wash in tap water
11. Counterstain in Mayer's haematoxylin blue for 1 min and wash in tap water
12. Dehydrate and clear mount

naturation step has been adapted from feulgen staining procedures and is more vigorous than that used in FCM, in which stoichiometry of propidium iodide intercalation into DNA must be maintained. A different monoclonal antibody has been used only to accommodate the avidin biotin method which is supplied as anti-mouse determinants.

Figure 7.5 shows two examples of squamous cell carcinomas from the head and neck region stained for BUdR incorporation using the above protocol. In both tumours, darkly stained cells which have incorporated BUdR are clearly identifiable. The intensity of the diaminobenzidine reaction has been maximised so as to detect efficiently all cells that have incorporated BUdR. The finer structure of BUdR incorporation can be obtained using less complete denaturation or lower antibody dilutions at the expense of compromising maximal detection of incorporated cells. There is a wide range of staining intensity in human tumours; some of this reflects cells which may have divided and shared their BUdR between the daughter progeny in the time between injection and surgical removal of the tumour; it will also reflect differences in the rate of DNA synthesis of cells in different parts of S phase when exposed to the "pulse" label.

Cellular differentiation of tumours has for many years been regarded as an important feature relating the natural history and response to treatment (Broders 1940). In oncology, the basic approach is to recognise that poorly differentiated tumours tend to have a shorter history and grow more rapidly than differentiated tumours. In response to radiotherapy and chemotherapy, they tend to regress in bulk more quickly and they show distant metastases sooner and more frequently than do the well-differentiated tumours (Kemp and Hendrickson 1982). It has been suggested that this may reflect differences in proliferative characteristics. This expectation has not proven to be true on the basis of our proliferation measurements using BUdR. The examples high-

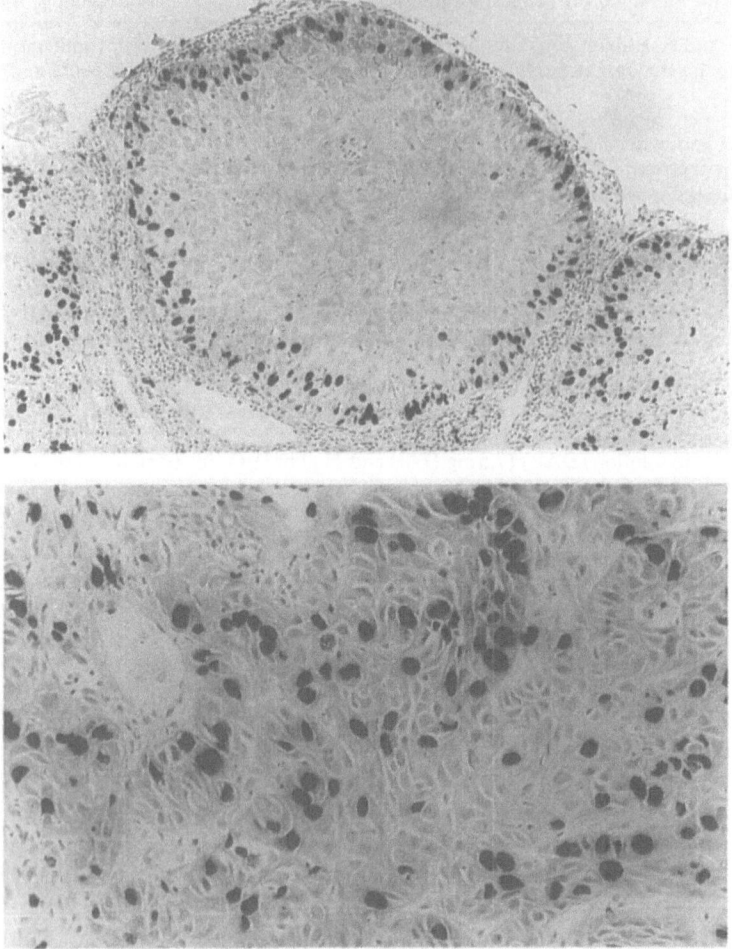

Figure 7.5. Immunohistochemical localisation of BUdR incorporation in squamous cell cancer of the head and neck. *Upper:* the marginal location of darkly staining BUdR labelled cells in the basal and transit cell layers of a downward projecting finger of a highly differentiated verruccous carcinoma of the tongue (×100). *Lower:* a more random staining pattern obtained in a poorly differentiated squamous cell carcinoma of the columella (×200).

lighted in Fig. 7.5 represent opposite ends of the differentiation spectrum. Fig. 7.5 (upper) demonstrates the BUdR localisation of a verrucous carcinoma showing the papilliferous structure of well-differentiated squamous epithelium penetrating into surrounding tissues. It is often quite difficult for the pathologist to find sufficient atypia to justify a malignant classification for the tumour; clinical features often have to be taken into consideration in order to come to the correct diagnosis. It might be expected that such well-differentiated tumours would be slow in their growth, characterised by a low LI and long cell cycle. This has proved not to be the case. In Fig. 7.5 the majority of cells in the basal and suprabasal layers show uptake of BUdR, suggesting that the growth

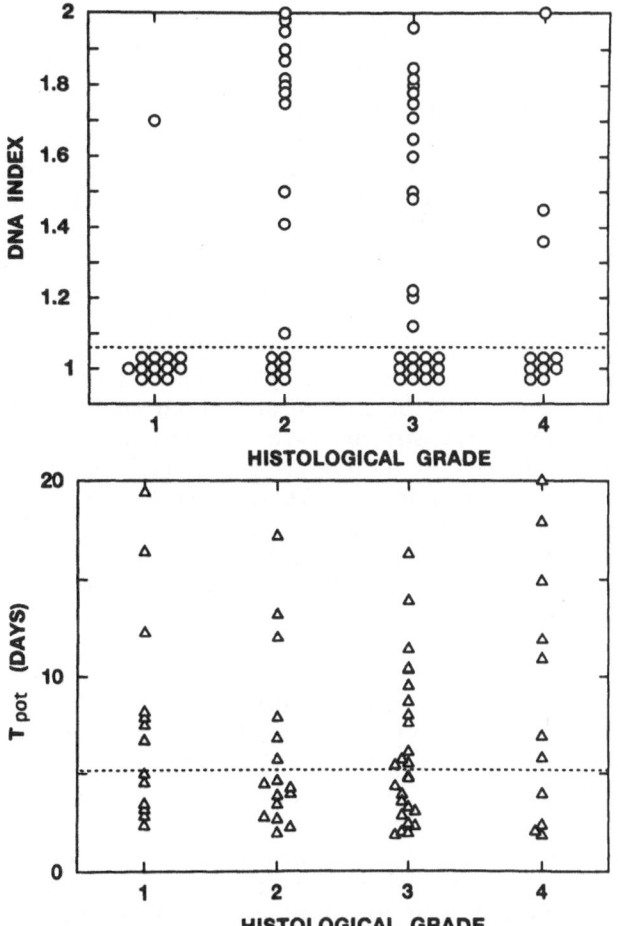

Figure 7.6. Relationship between histopathological grading and DNA index (*upper*) or T_{pot} (*lower*) in 68 squamous cell carcinomas of the head and neck region. *Upper:* dotted line denotes the delineation of diploid from aneuploid tumours (for clarification, some diploid tumours have been ascribed higher or lower DNA indices). *Lower:* dotted line represents the median T_{pot} value of this particular group of tumours.

fraction in the actively reproducing cells is very high. Thus, although the overall growth rate of the tumour may not be high due to maturation and exfoliation, cell production rates will far exceed those in normal squamous mucosa.

Fig. 7.5 (lower) demonstrates the opposite type of squamous cell tumour. This was a poorly differentiated infiltrating carcinoma of the columella. The staining pattern shows little evidence of a marginal localisation and is more random in nature. The interesting point to note was that proliferation, measured by FCM, was more rapid (T_{pot} = 5.1 days) in the verruccous carcinoma than it was in the poorly differentiated columella carcinoma (T_{pot} = 10.3 days). In certain sites, we have found consistently that proliferation is independent of histology (Dische et al. 1989).

From the results obtained from 68 SCC tumours of the head and neck region classified according to the criteria of Broders (1940) and correlated to either the presence of DNA aneuploidy or T_{pot} (Fig. 7.6), it can be seen that a relationship exists between grading and the presence of abnormal DNA stemlines. Only 1 of 13 (8%) grade 1 tumours was aneuploid, whilst in grade 2 and 3 tumours the incidence of aneuploidy was 67% and 54% respectively. Interestingly, only 3 of 11 (27%) of grade 4 tumours were aneuploid. In contrast, Fig. 7.6b shows that there is an almost equal distribution of rapidly and slowly proliferating tumours in each histological grade. The median value for T_{pot} was 5.2 days overall, whilst in grades 1, 2, 3 and 4, the median values were 6.7, 4.0, 5.1 and 6.9 days, respectively. This relationship has also been shown for colorectal adenocarcinomas (Rew et al. 1991), in which T_{pot} was independent of histopathological differentiation or Dukes' classification. The lack of correlation between proliferation and grading is not the case in all tumour sites; strong relationships have been found in breast, gastric and certain other tumour types.

Histological quantification of BUdR localisation is an integral aspect of the proliferative classification of human tumours. We are currently assessing several parameters. Firstly, the pathologist can estimate the percentage of the specimen occupied by tumour, normal stromal elements and necrosis in each tumour. This information can be used to correct the FCM-derived information in diploid tumours where discrimination based on DNA content is not possible. The pathologist can quantify the overall labelling index in the tumour components as well as maximum and minimum LI. In the two tumours illustrated in Fig. 7.5, the overall LI was 10% and 6% for the verruccous and columella tumour, respectively, yet areas of 20% and 50% cell labelling were seen in each of these tumours. One other aspect which is proving to be of interest is the scoring of BUdR-labelled mitotic cells. In most tumours, enough time has elapsed between injection and biopsy to observe BUdR-labelled cells in G_1 (these must have divided). If all cells were in cycle, then all mitoses should be labelled. This is not always the case; the incidence of BUdR-labelled mitoses ranges from as low as 10% of all mitoses up to almost 100%. This would suggest the presence of slowly cycling or quiescent cells, or it may be an indication of cell loss through aborted mitosis. The quantification of histological localisation of BUdR in the human tumours we have studied is still in an early stage, but it already appears to be an important adjunct to the information generated by FCM. The combination of these two techniques provides a picture of the proliferative nature of human solid tumours and normal tissues that is not obtainable by any other method.

Clinical Correlations

The test of any prognostic indicator is whether it is important enough to play a role in determining which treatment modality or schedule is the most appropriate for a particular patient or group of patients. In general, proliferation measurements have failed at this final hurdle. There is no doubt that proliferation is an important parameter determining the clinical outcome for a variety of tumours (Tubiana and Courdi 1989), but measurements have yet to influence treatment (see Chap. 10). Much of this has been due to the methods used to measure proliferation, in particular ^{3}H-thymidine autoradiography, which is too slow to be used prospectively. The advent of BUdR-FCM has allowed the opportunity for rapid estimation of potentially better measures of proliferation. Radiotherapists have been amongst the first to exploit this new opportunity. The development of BUdR-FCM has coincided with a growing awareness amongst radiotherapists that conventional fractionation may be failing to cure some patients due to rapid proliferation or repopulation of tumours during radiotherapy (Amdur et al. 1989; Taylor et al. 1990; Withers et al. 1988). Analysis of split-course radiotherapy schedules incorporating a week or 2 weeks without treatment have shown that clonogenic cells repopulate at a rate similar to T_{pot} measured by the BUdR-FCM method (Fowler 1986). The values for the effective doubling times of clonogenic cells have been calculated by making assumptions about the loss of local control or the extra dose required to achieve the same local control compared to a continuous course of radiotherapy.

If repopulation of tumours is a problem, the logical approach to control proliferation would be to shorten the overall treatment time by giving two or three fractions per day.

At Mount Vernon Hospital, a radiotherapy schedule known as CHART (continuous, hyperfractionated, accelerated radiotherapy) has been pioneered. This consists of administering three fractions of 1.5 Gray per day for 12 continuous days (Saunders et al. 1986).

We have made a preliminary evaluation of patients treated in the pilot study of CHART, whose tumour cell kinetics were assessed by BUdR administration and FCM. The local tumour control within a subset of 38 CHART patients all of whom had primary SCC tumours in the head and neck region is shown in Fig. 7.7. This particular subset of head and neck tumours was more aggressive than the group as a whole, as they were chosen for accessibility of biopsy. This is highlighted by the median values for LI, T_S and T_{pot}, which were 7.9%, 9.8 h and 3.9 days, respectively. The data are presented for each proliferation parameter using a cut-off at the appropriate median value. Of the 38 patients, 28 (74%) achieved complete local tumour control, subsequently five of these tumours recurred in the treatment field.

It can be seen that there was no significant influence of any of the proliferation parameters on local tumour control. This result is expected, since the accelerated treatment was designed to overcome the problem of tumour cell repopulation during treatment. It is in a "conventional" arm of treatment that one would expect to see tumours with a short T_{pot} dominating in the failure group.

The pilot study of CHART has led to a multi-centre randomised controlled trial of CHART versus conventional fractionation. Measurements of T_{pot} are being made in this trial. The important results will emerge from the conventional fractionation where it is envisaged that tumours with short T_{pot} values will do badly.

Preliminary data from an EORTC-sponsored trial of accelerated (5 weeks) versus conventional (7 weeks) radiotherapy (Begg et al. 1991), in which T_{pot} measurements are also being made, support the hypothesis outlined above. The local tumour control data for the accelerated and conventional arm has been classified as fast or slow using a cut-off of 4.0 days, which was close to the median value. Fig. 7.8 demonstrates the loss of local tumour control in fast-proliferating tumours treated with conventional fractionation. These data

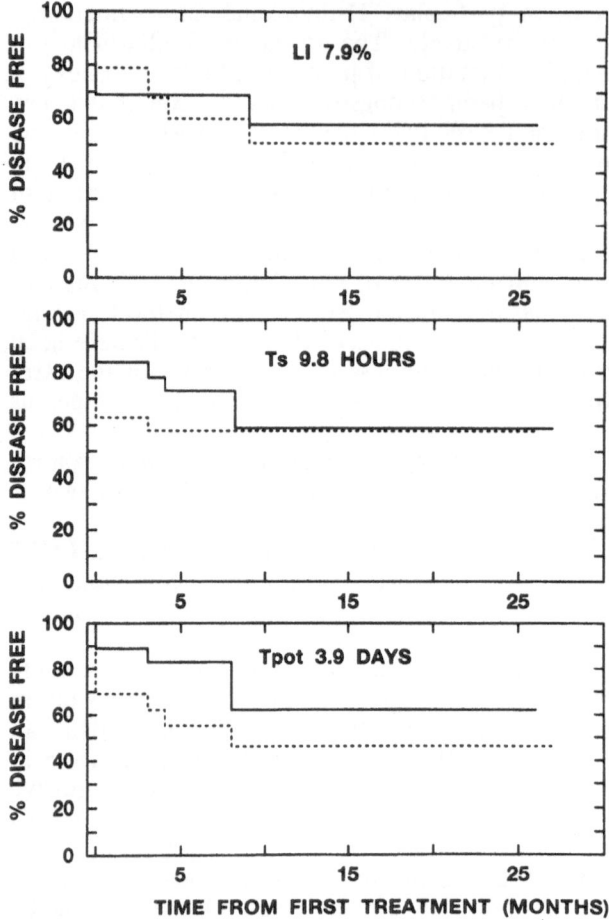

Figure 7.7. The influence of proliferation parameters on local tumour control in 38 patients with primary cancer of the head and neck treated with CHART. The data have been divided according to above (——) or below (-----) the median value of each parameter as indicated in each panel. Using a variety of statistical tests, none of the proliferation parameters showed any significant relationship with local tumour control.

also suggest, as demonstrated with CHART, that acceleration of treatment overcomes proliferation such that fast and slow tumours show similar responses.

The data generated from the CHART and EORTC trial should establish the importance of pre-treatment T_{pot} measurements in the management of radiotherapy patients.

Conclusion

We have described the application of BUdR labelling and flow cytometry to the study of cell proliferation in human tumours. In terms of versatility, ease

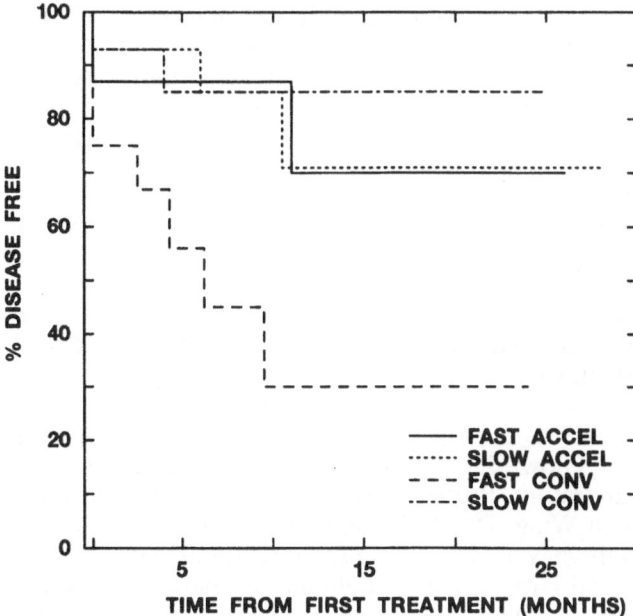

Figure 7.8. The influence of proliferation on local tumour control in patients treated with an EORTC accelerated radiotherapy regime, compared with conventional fractionation. Life tables have been divided as fast or slow proliferation in either the accelerated (accel.) or conventional (conv.) arm. In the conventional arm, there is a significant difference between rapid and slow tumours ($P = 0.02$ from χ^2 test). (From Dr A.C. Begg, with permission.)

of analysis, and lack of toxicity, there is no doubt that this technology has opened up new and exciting areas of study, particularly of human tumours and normal tissues, in ways that would never be possible with conventional auto-radiography. Indeed, when combined with immunohistochemical localisation of BUdR (or IUdR) labelled cells in tissue sections, the techniques we have described can do all that autoradiography can do and more. The use of surgical resection can allow studies to be made on adjacent normal tissues (e.g. colorectal tissue).

Under appropriate conditions, BUdR labelling can be combined with markers for cell surface antigens, oncogene products, cell cycle proteins, etc. to provide further information on the control of proliferation in human tumours and tissues.

As regards the specific applications that we have described, the BUdR technique can now be regarded as essentially routine. Apart from the necessity for a biopsy that may not be needed for diagnosis, the procedures are quite straightforward and non-toxic. As current prospective studies of the predictive value of the pre-treatment T_{pot} are concluded in the next year or so, we believe that its value will be clearly established in characterising tumours prior to radiotherapy and this should also encourage wider applications in chemotherapy studies.

Experimentally, the BUdR technique has lead to a resurgence of interest in the general field of cell kinetics, expanding the types of study that can be done with greater precision, allowing more detailed studies of slowly proliferating normal tissues, for instance, to be made (Wilson et al. 1987).

We have not discussed in great detail techniques of tissue disaggregation. Nevertheless, it is essential to remember that, with flow cytometry, tissue architecture is lost and the technique may be subject to sampling artefacts. It is for these reasons that we believe that all samples analysed by flow cytometry must have adjacent material fixed in formalin for histological analysis.

In most of the cases when we have failed to obtain good FCM information, the sections have shown either that there was little or no tumour present or that it was essentially necrotic.

Finally, as in all flow cytometry, one should remember Shapiro's maxim: "Garbage in, garbage out!" Scrupulous attention to detail at all stages of tissue preparation will ensure a high success rate.

Acknowledgements. This work is supported by the Cancer Research Campaign. We would like to acknowledge the support of our colleagues, Chris Martindale at the Gray Laboratory, Professor S. Dische and Dr M.I. Saunders at the Marie Curie Research Wing, Dr M.H. Bennett and Mrs A. O'Halloran at the Department of Pathology at Mount Vernon Hospital and Mr D.A. Rew at Southampton General Hospital. We would like to thank Mrs Dorothy Brown for preparation of the manuscript.

References

Amdur RJ, Parsons JT, Mendenhall WM, Million RR and Cassisi NJ (1989) Split-course *versus* continuous-course irradiation in the post-operative setting for squamous cell carcinoma of the head and neck. Int J Radiat Oncol Biol Phys 17:279–285

Begg AC (1989) Derivation of cell kinetic parameters from human tumours after labelling with bromodeoxyuridine or iododeoxyuridine. In: The scientific basis of modern radiotherapy. British Institute of Radiology Report 19:113–119

Begg AC, McNally NJ, Shrieve DC and Karcher H (1985) A method to measure the duration of DNA synthesis and the potential doubling time from a single sample. Cytometry 6:620–626

Begg AC, Hofland I, Moonen L et al. (1991) The predictive value of cell kinetic measurements in a European trial of accelerated fractionation in advanced head and neck tumours: an interim report. Int J Radiat Oncol Biol Phys (in press)

Broders AC (1940) The microscopic grading of cancer. In: Pack GT, Livingston EM (eds) The treatment of cancer and allied diseases, vol 1. Paul B. Hoeber, New York, pp 19–41

Dische S, Saunders MI, Bennett MH, Wilson GD and McNally NJ (1989) Cell proliferation and differentiation in squamous cancer. Radiother Oncol 15:19–23

Dolbeare F, Gratzner HG, Pallavicini M and Gray JW (1983) Flow cytometric measurement of total DNA content and incorporated bromodeoxyuridine. Proc Soc Natl Acad Sci 80:5573–5577

Fowler JF (1986) Potential for increasing the differential response between tumours and normal tissues: Can proliferation rate be used? Int J Radiat Oncol Biol Phys 12:641–645

Gratzner HG (1982) Monoclonal antibody to 5-bromo and 5-iododeoxyuridine: a new reagent for detection of DNA replication. Science 218:474–475

Kemp RL and Hendickson MR (1982) The impact of pathology on cancer treatment. In: Carter SK, Gladstein E and Livingston RL (eds) Principles of cancer treatment. McGraw-Hill Book Co., New York, pp 26–33

Meyer JS (1982) Cell kinetic measurements of human tumours. Hum Pathol 13:874–877

Rew DA, Wilson GD, Taylor I and Weaver PC (1991) Proliferation characteristics of human colorectal carcinomas measured *in vivo*. Br J Surg 78:60–66

Saunders MI and Dische S (1986) Radiotherapy employing three fractions in each day over a continuous period of 12 days. Br J Radiol 59:523–525

Saunders MI, Dische S, Fowler JF et al. (1988) Radiotherapy employing three fractions on each of twelve consecutive days. Acta Oncol 27:163–167

Steel GG (1977) Growth kinetics of tumours. Clarendon Press, Oxford

Taylor JMG, Withers HR and Mendenhall WM (1990) Dose-time considerations of head and neck squamous cell carcinomas treated with irradiation. Radiother Oncol 17:95–102

Tubiana M and Courdi A (1989) Cell proliferation kinetics in human solid tumours; relation to probability of metastatic dissemination and long term survival. Radiother Oncol 15:1–18

White RA (1989) Computing multiple cell kinetic properties from a single time point. J Theor Biol 141:429–446

White RA and Meistrich ML (1986) A comment on "A method to measure the duration of DNA synthesis and the potential doubling time from a single sample". Cytometry 7:486–490

White RA, Terry NHA, Meistrich ML and Calkins DP (1990) Improved method of computing potential doubling time from flow cytometric data. Cytometry 11:314–317

Wilson GD, McNally NJ and Dische S, et al. (1988) Measurement of cell kinetics in human tumours *in vivo* using bromodeoxyuridine incorporation and flow cytometry. Br J Cancer 58:423–431

Wilson GD, Soranson JA and Lewis AA (1987) Cell kinetics of mouse kidney using bromodeoxyuridine incorporation and flow cytometry: preparation and staining. Cell Tissue Kinet 20:125–133

Withers HR, Taylor JMG and Maciejewski B (1988) The hazard of accelerated tumour clonogen repopulation during radiotherapy. Acta Oncol 27:131–146

8 The Application of Immunohistochemistry in Assessment of Cellular Proliferation

C.C.-W. YU, A.L. WOODS and D.A. LEVISON

Introduction

The use of immunohistochemical techniques enables the visualisation of specific cell cycle related antigens in tissue, with the important advantage of preserving the spatial orientation of cells. Antibodies may recognise endogenous cell cycle related molecules occurring naturally within the tissue to be studied, or exogenous substances, such as bromodeoxyuridine (BUdR), a thymidine analogue incorporated during the S (DNA synthetic) phase of the cell cycle, which may be introduced by in vivo administration (see Chap. 7) or in vitro incubation. There is a wide range of targets for these antibodies, most of which are cell cycle associated antigens, but it is also possible that markers of quiescent and non-proliferating cells and markers of transition from the quiescent to the proliferating state may convey information about proliferative activity (Hall and Woods 1990).

Cell Cycle Associated Antigens

Cell cycle associated antigens may be localised in the nucleus, the cytoplasm or the cell membrane. They may be associated with all phases of the cell cycle or with a particular phase, for example, mitosis. The list of potential targets for antibodies is as vast as our present knowledge of the cell cycle and its regulatory mechanisms will allow. Immunostaining with antibodies to some target molecules not directly associated with the cell cycle may also demonstrate a relationship with established proliferative indices, for example recent data from D.M. Barnes et al. (personal communication) have shown that those cases of breast cancer with high expression of the receptor tyrosine kinase proto-oncogene c-*erb*-B-2 tend to have high proliferative indices as assessed by thymidine incorporation. However, some cases with high thymidine incorporation indices do not express c-*erb*-B-2, perhaps as a consequence of the numerous pathways to mitogenesis in cells. The antibodies which will be described in detail are reported antibodies that appear to have the most potential in assessing cellular proliferation; the most extensively evaluated of these are Ki-67 and antibodies to proliferating cell nuclear antigen.

Ki-67 Antigen

Ki-67 is a mouse monoclonal antibody that identifies a poorly characterised nuclear antigen (see Fig. 8.1) associated with the cell cycle (Gerdes et al. 1983). Detailed cell cycle analysis has demonstrated that the Ki-67 antigen is expressed in all phases except G_0 and early G_1 (Gerdes et al. 1984). The amount of antigen varies throughout the cell cycle and reaches a maximum during G_2 and M phase (Sasaki et al. 1987; Wersto et al. 1988). Recent research suggests that the antigen is a component of the nuclear matrix (Verheijen et al. 1989a, b), and the gene encoding it has recently been cloned. Numerous studies have shown a good correlation between Ki-67 immunore-activity and other indices of cell proliferation, such as flow cytometry (Baisch and Gerdes 1987; Schwarting et al. 1986; Walker and Camplejohn 1988), thymidine labelling (Kamel et al. 1989; Schwarting et al. 1986; Silvestrini et al. 1988; Veroni et al. 1988) and BUdR incorporation (Falini et al. 1988; Sasaki et al. 1988; Schrape et al. 1987; Silvestrini et al. 1988; Wersto et al. 1988). However, recent evidence suggests that the fraction of cells showing Ki-67 immunoreactivity consistently overestimates the growth fraction of tumours (Scott et al. 1991).

Nevertheless, staining with Ki-67 has been widely used as an operational marker of cell proliferation and has been studied in many neoplastic and non-neoplastic conditions. It seems to provide prognostically useful information in non-Hodgkin's lymphoma (Hall et al. 1988a), since a relatively high Ki-67 index appears to be associated with reduced survival in low-grade lymphomas. Theoretically, this can identify a group of patients with poor prognosis, and allow them to receive more aggressive initial treatment in the hope of producing improved survival rates. In contrast, a very high Ki-67 index is associated with an improved prognosis in high-grade lymphomas. This latter observation is consistent with clinical observations that rapidly growing tumours often respond well to chemotherapy. A likely explanation is that, in high-grade lymphomas with less high Ki-67 indices, chemotherapy fails to eliminate some cells which are in the resting phase of the cell cycle (see Chap. 11). In breast carcinoma, Ki-67 results have been less encouraging. Walker and Camplejohn (1988) have compared Ki-67 immunostaining, flow cytometric analysis and histological grade in breast carcinoma, but have failed to demonstrate a reli-able correlation between these parameters. Similarly, Shepherd et al. (1988) found no correlation between Ki-67 positivity and known prognostic para-meters in a series of colorectal carcinomas.

The results of the various reported studies using Ki-67 have been sum-marised in a recent review (Brown and Gatter 1990). The applications of this antibody have also been investigated in non-neoplastic conditions: a study of patients with ulcerative colitis showed a higher degree of positivity in patients with active compared with inactive disease, suggesting that Ki-67 labelling may provide an independent marker of disease activity (Franklin et al. 1985).

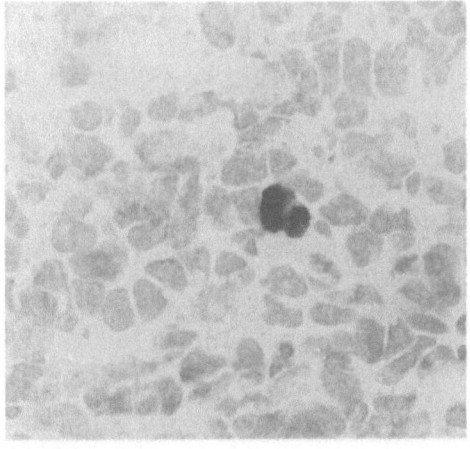

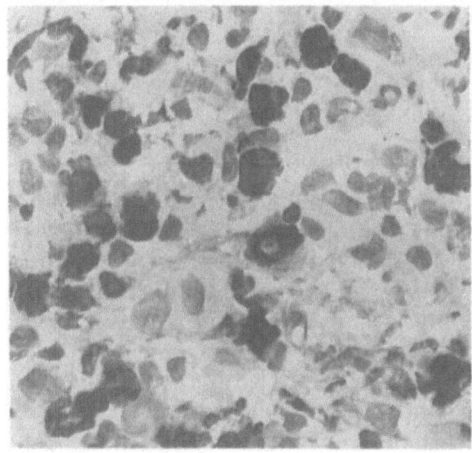

Figure 8.1. Examples of immunostaining with the antibody Ki-67 on frozen sections, showing a low-grade non-Hodgkin's lymphoma (*upper*) with a low proportion of positive cells and a high-grade non-Hodgkin's lymphoma (*lower*) with a high proportion of positive cells.

However, a significant drawback to the use of this antibody is the need for fresh or snap-frozen material, as the antigen is very sensitive to fixation.

Proliferating Cell Nuclear Antigen

Proliferating cell nuclear antigen (PCNA) is an evolutionarily highly conserved 36 kDa acidic nuclear protein, essential for DNA synthesis. It can be detected in proliferating cells by autoantibodies in serum from patients with systemic lupus erythematosus (Miyachi et al. 1978; Takasaki et al. 1987). PCNA functions as an auxiliary protein for DNA polymerase δ in DNA synthesis (Bravo et al. 1987; Fairman 1990; Prelich et al. 1987; Tan et al. 1986). However, it may also be involved in unscheduled DNA synthesis (Ogata et al. 1987; Suzuka et al. 1989). Immunofluorescent studies of cultured cells have indicated that PCNA is a cell-cycle related antigen (Celis and Celis 1985). These studies have suggested that there are two populations of PCNA present during S phase (Bravo and Macdonald-Bravo 1987). Nucleoplasmic PCNA corresponds to the

PCNA present at low levels in quiescent cells that are capable of cell division, and is not apparent in cells fixed in organic solvents such as methacarn. The second form of PCNA is associated with DNA replication sites and cannot be extracted with organic solvents.

Recent data have shown that there is only a two- to three-fold increase in the total level of PCNA during the cell cycle, but there is a marked increase in the proportion of replication site-associated PCNA (Morris and Mathews 1989). The quantity of PCNA present in the nucleus of cycling cells is greater than that required for DNA synthesis (Morris and Mathews 1989). The half-life of PCNA is approximately 20 h, and PCNA has been detected by immunological methods in cells that have recently left the cell cycle. However, it is undetectable in long-term quiescent cells using the above methods.

The gene for PCNA is transcribed in both quiescent and proliferating cells, but PCNA mRNA only accumulates in proliferating cells (Chang et al. 1990). In quiescent cells, PCNA mRNA is unstable and, in this state, intron 4 is present within the transcript. If this intron is removed, high levels of PCNA mRNA accumulate (Ottavio et al. 1990). It has been shown that growth factors, especially platelet-derived growth factor (PDGF), increase the stability of PCNA mRNA and allow PCNA translation. Accumulation of PCNA protein is not inevitably associated with DNA synthesis. In fact, PCNA will accumulate in the presence of hydroxyurea, which inhibits DNA synthesis (Bravo and Macdonald-Bravo 1984, 1985; Jaskulski et al. 1988). Consequently, the exact role of PCNA in DNA replication is unclear, DNA polymerase δ has been implicated in a PCNA-independent DNA repair process (Nishida et al. 1988; Syraoja and Linn 1989). However, PCNA may be involved in this process, since redistribution of PCNA to sites of DNA damage after ultraviolet irradiation has been reported in non-S-phase cells (Celis and Madson 1986; Toschi and Bravo 1988).

Recently, monoclonal antibodies have been generated to genetically engineered PCNA (Waseem and Lane 1990) and one of these, designated PC10, demonstrates the proliferative compartment in conventionally fixed and processed normal human tissues (Fig. 8.2; Hall et al. 1990). PCNA immunoreactivity can be demonstrated after fixation in a wide range of solutions, including formalin, methacarn and Bouin's reagent. In all cases except methacarn, both diffuse and granular nuclear staining can be identified; with methacarn fixation, only granular staining is present. The significance of this is unclear. The interpretation of PCNA immunoreactivity necessitates careful consideration of the length of fixation of the tissues and the mode of preparation (Hall et al. 1990). It has been shown that PCNA immunoreactivity in non-neoplastic tissues is greatly reduced after 48 h of fixation in formalin, and

---

Figure 8.2. Normal formalin-fixed paraffin-embedded tissues immunostained with PC10 (anti-PCNA). *Upper left*: Tonsil under a low-power objective showing PCNA immunoreactivity concentrated in germinal centres (clearly demonstrating zoning), but also scattered positive cells in the interfollicular areas. *Upper right*: Jejunal mucosa, with the highest proportion of stained cells in the crypts and progressive loss of immunoreactivity on going up the villi. *Lower*: Normal proliferative-phase endometrium (*lower left*) showing positive epithelial and stromal cells, contrasting with secretory phase endometrium (*lower right*) with no positive cells in this field.

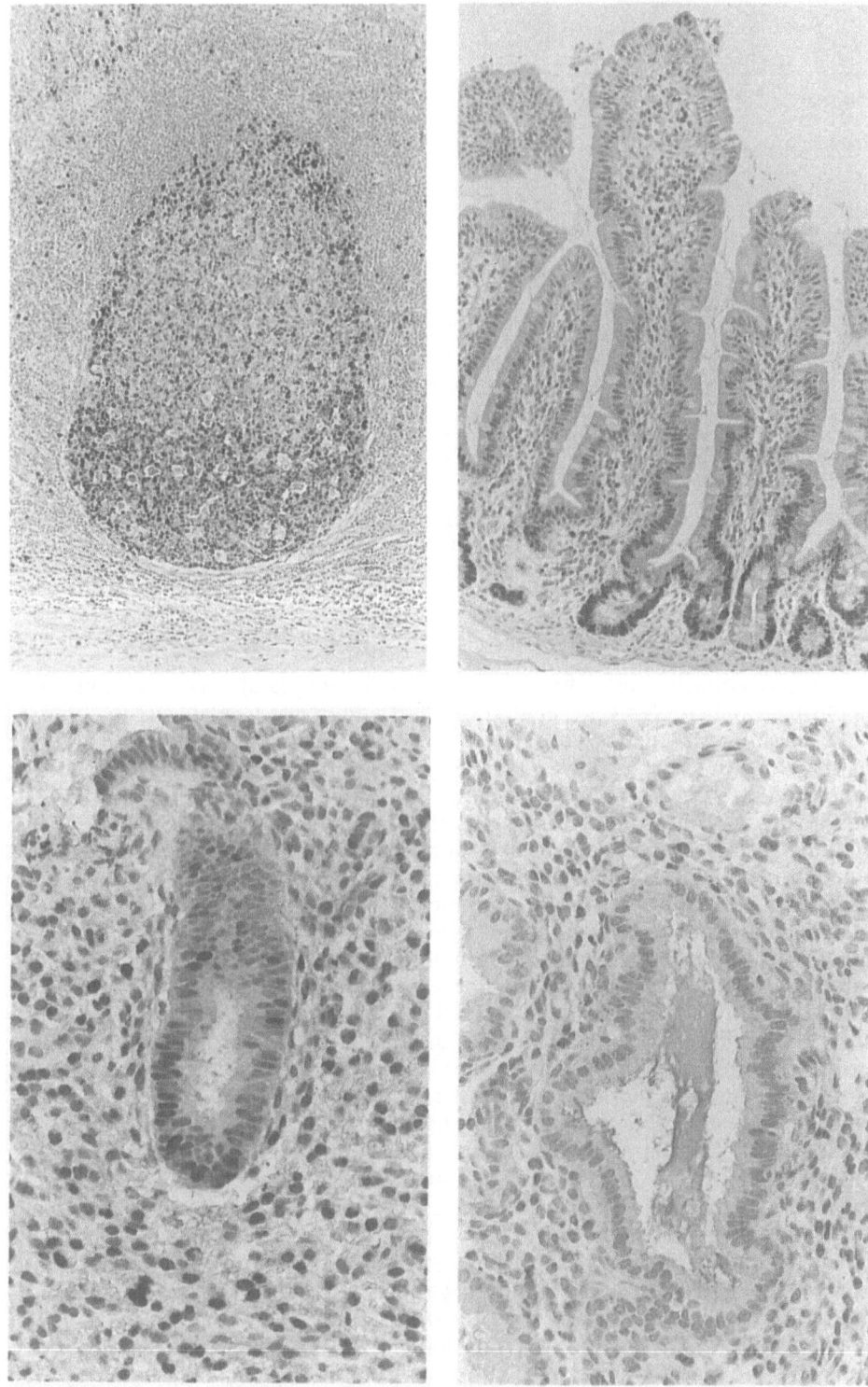

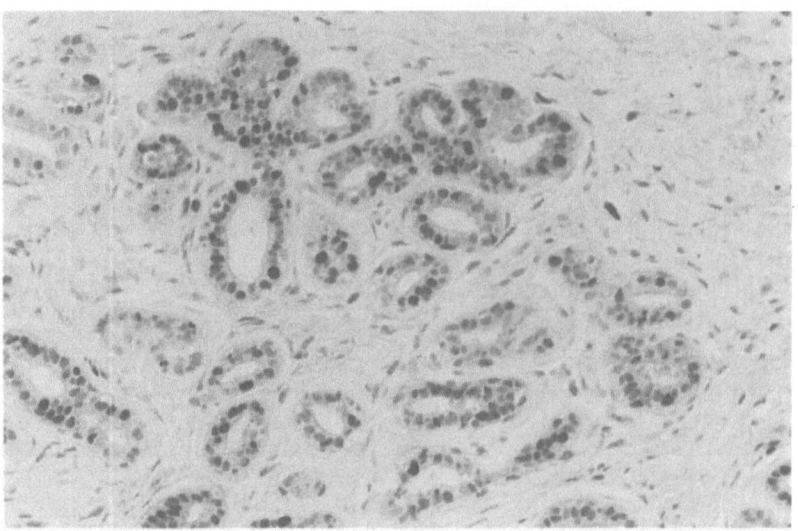

Figure 8.3. PC10 immunostaining of a histologically normal breast lobule showing many positive nuclei. Adjacent to this field was an infiltrating ductal carcinoma. There are considerably more PCNA immunoreactive cells than usually observed in the absence of a neoplasm.

is virtually abolished after fixation for 72 h. Staining is also greatly reduced or abolished if cut sections are heated to assist adherence to glass slides.

PC10 immunoreactivity has been shown to correlate with Ki-67 staining in nodal lymphomas (Hall et al. 1990) and with $S+G_2M$ phase fraction as measured by flow cytometry in gastrointestinal lymphomas (Woods et al. 1991). In gastrointestinal lymphomas, the proportion of cells showing positive staining with PC10 correlates well with histological grading and with prognosis: tumours with high counts tend to be high-grade lesions with a relatively poor prognosis. In addition, PC10 immunoreactivity can be induced in phyto-haemagglutinin-stimulated human peripheral blood mononuclear cells in parallel with bromodeoxyuridine labelling. After induction of macrophage differentiation by phorbol esters, there is a similar reduction in staining of HL60 cells with PC10 and Ki-67 (Hall et al. 1990). The above data suggest that PC10 immunoreactivity may be a marker of cell proliferation in fixed histological material. However, studies of gastric carcinoma, breast carcinoma and haemangiopericytomas (a soft tissue tumour) suggest that there is a poor correlation with other indices of proliferation in these particular types of neoplasm (D.M. Barnes et al. unpublished; Jain et al. 1991; Yu et al. 1991). In these situations, more cells stain with PC10 than is expected; possible explanations include deregulated expression of PC10 in malignant lesions and the long half-life of PCNA, leading to its persistent expression in some cells which are not actively dividing (Hall et al. 1990). In spite of this, immuno-staining for PCNA in haemangiopericytomas appears to give an indication of clinical outcome: patients who had recurrent tumours and metastases and who died rapidly from malignancy were found to have a high proportion of positive cells in their tumours (Fig. 8.4; Yu et al. 1991). Another study has suggested

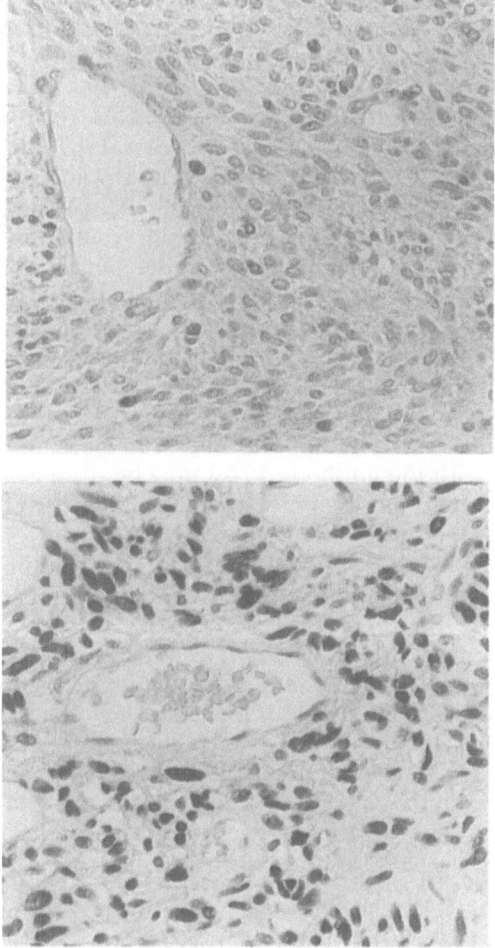

Figure 8.4. PC10 immunostaining of haemangiopericytomas. *Upper*: Benign haemangiopericytoma with only occasional cells showing positive nuclear staining. *Lower*: Malignant haemangiopericytoma with most of the cells in this field showing positive nuclear staining.

that immunostaining with PC10, assessed by a semi-quantitative method, may also be of prognostic value in gastric carcinoma (Jain et al. 1991).

An additional interesting observation is that in histopathologically normal tissues adjacent to some breast, gastric and pancreatic tumours, a marked increase in PCNA immunostaining is seen (see Fig. 8.3). It may be that this appearance is caused by the production of growth factors, e.g. PDGF, by the tumours, this in turn inducing PCNA mRNA stability and, consequently, PCNA expression. Nevertheless, the results of staining on normal tissues and lymphomas suggest that PCNA immunostaining may be employed as a reliable marker of cell proliferation in non-neoplastic tissues and in lymphomatous lesions. Its usefulness and the biological significance of PCNA expression in non-lymphoid neoplasms requires further evaluation.

The antibody PC10, which has been described in detail above, produces strong staining with tissue fixed in formalin, the fixative most commonly used for routine diagnostic histopathology, as well as a range of other fixatives. Other antibodies have been developed to PCNA; the one which has been reported most extensively in the literature, named 19A2, produces readily detectable staining in methanol-fixed tissues (Galand and Degraef 1989) but no, or only faint, nuclear staining in material fixed in ethanol, Bouin's reagent, Carnoy's or formaldehyde. However, hydrolysis with 2N HCl prior to immunocytochemistry has been reported to somewhat improve the results with Bouin's or Carnoy's and markedly augment the intensity of the peroxidase reactions in formaldehyde- and in methanol-fixed tissues. Double labelling with tritiated thymidine and with the PCNA antibody revealed that, in methanol-fixed tissues, the PCNA labelling index did not differ by more than 6% from the tritiated thymidine index. Other workers have found a good correlation between thymidine labelling index and PCNA labelling index in normal breast tissue (Battersby and Anderson 1990). Thus it appears that, in methanol-fixed tissues, PCNA immunostaining with the antibody 19A2 may be used as a substitute for the tritiated thymidine autoradiographic labelling index. From this, Galand and Degraef (1989) have proposed a method for measuring S-phase duration in tissues fixed at a known time interval after a single labelling with tritiated thymidine followed by PCNA immunocytochemical detection and autoradiography.

Other antibodies to PCNA have not been fully assessed as yet but, so far, research with PC10 and 19A2 suggests that PCNA is a target for immunohistochemical staining, which may be a practical method for assessing cellular proliferation in clinical material. The main advantage of these antibodies over the others described, is that they do not require fresh or frozen tissue, and therefore retrospective studies can be carried out on routine paraffin-embedded material. It should be noted that there may be important differences between antibodies that recognise distinct epitopes on the PCNA molecule.

DNA Polymerase δ and α

Monoclonal antibodies have been described to DNA polymerase δ, a mammalian DNA polymerase with 3' to 5' exonuclease activity (Lee et al. 1989). Using immunochemical methods, Marraccino et al. (1987) have shown that there is cell cycle alteration in the level of DNA polymerase δ with peaks during G_2 and M phases. Whether antibodies to DNA polymerase δ could be of value in assessing cell proliferation remains to be determined. However, recent studies suggest that DNA polymerase δ is temporally closely linked to DNA polymerase α (Lee et al. 1989).

DNA polymerase α is recognised to be a key enzyme in eukaryotic DNA synthesis, and recent studies have demonstrated that, with transition from G_0 to G_1, the steady state mRNA levels, rate of synthesis of new DNA polymerase α and the in vitro enzymatic activity exhibit a large and concordant increase occurring before the peak of DNA synthesis (Wahl et al. 1989). Monoclonal antibodies that recognise this key DNA polymerase have been reported (Tanaka et al. 1982). Immunocytochemical studies of tumour cell lines have shown diffuse nuclear staining with sparing of the nucleolus in the majority of cells (Bensch et al. 1982). In human diploid fibroblasts, a smaller proportion of cells show nuclear staining and those cells that have become contact-inhibited fail to stain. During mitosis, with dissolution of the nuclear membrane, staining is distributed throughout the cell. Anti-DNA polymerase α monoclonal antibodies have been applied to cryostat sections of a number of normal and pathological tissues (Mushika et al. 1989; Namikawa et al. 1987a, b; Yamaguchi et al. 1990), providing information concerning the number and distribution of proliferating cells. Unfortunately, all of the antibodies reported so far only work on frozen sections.

p125/6.5 Antigen

Philipova et al. (1987) have reported a monoclonal antibody that recognises a 125 kDa protein with a pI of 6.5, hence known as p125/6.5, associated with the nuclear matrix in proliferating cells. It is not present in resting lymphocytes, but is induced on phytohaemagglutinin stimulation in parallel with DNA synthesis. Two-dimensional gel electrophoresis and immunoblotting studies indicate that the antigen increases in mitotic cells, compared with interphase cells, and this is consistent with immunofluorescence studies indicating maximal fluorescence during metaphase and anaphase (Todorov et al. 1988).

The p125/6.5 antigen is present in a variable proportion of cells in a range of tumours and it is seen in normal tissues in a fraction of cells that approximates to those cells known to be cycling, as determined by more conventional methods such as tritiated thymidine incorporation (Yankulov et al. 1989). Although a detailed assessment of normal tissues has not yet been reported, a recent immunocytochemical study of spermatogenesis (Hadjiolova et al. 1989) provides convincing evidence for the proposition that the p125/6.5 nuclear matrix antigen is characteristic both of proliferating cells and of those cells committed to proliferation. Recent unpublished results suggest that the antigen is immunologically detectable in non-cycling cells (B. Ansari and P.A. Hall, personal communication).

K 112 Antigen

The monoclonal antibody K 112 was generated after a single intrasplenic immunisation of mice with tissue from a recurrent laryngeal squamous cell carcinoma (Quak et al. 1990). The antibody detects a 43 kDa nuclear antigen, with a pI of 5.4, which is expressed only in cycling cells. Expression is typically seen in a granular pattern, excluding the nucleoli. The K 112 antigen remains detectable after in situ nuclease digestion of cells followed by high salt extraction, suggesting that it is associated with the nuclear matrix. During mitosis, the bulk of the antigen is diffusely distributed in the cytoplasm. Immunostaining has been obtained in a variety of cell lines and normal and neoplastic human tissues. However, the antibody has only been shown to work on frozen sections.

Transferrin Receptor

The cell cycle related expression of the transferrin receptor (CD71) has been used as an index of cellular proliferation. In non-Hodgkin's lymphoma, CD71 immunoreactivity has been shown to provide clinically relevant data by distinguishing low-grade from high-grade tumours (Habeshaw et al. 1983). However, the transferrin receptor is expressed widely, not only by proliferating tissues but also by metabolically active non-proliferating tissues (Gatter et al. 1983), and it is thus a poor index of cellular proliferation (Pileri et al. 1987).

Mitosis-Associated Changes

Light microscopic and ultrastructural studies indicate that there are considerable alterations in the structure of the nucleus during cell division (Alberts et al. 1989). Such morphological alterations are almost certainly associated with numerous biochemical alterations, some of which may potentially be detected by immunological methods. For example, the changes in histones may represent possible targets for antibody detection of cell cycle alterations: Chou et al. (1990) have reported that the demonstration of histone H3 mRNA by in situ hybridisation may be an effective method for demonstrating proliferating cells in histological material. The immunohistological demonstration of mitotic cells by a monoclonal antibody, C_5F_{10}, has been reported (Lloyd et al. 1985), but the antibody has been found to cross-react with many cellular components (Hall and Levison 1990) and does not seem to offer any advantage over conventional assessment of mitoses in histological sections.

Bromodeoxyuridine

Bromodeoxyuridine (BUdR) is a thymidine analogue incorporated during the S phase of the cell cycle. It may be introduced into fresh, unfixed tissue by in vitro incubation. Tissue penetration is enhanced if this process is performed at high pressures (typically three times atmospheric pressure). Various monoclonal antibodies to BUdR have been developed (Gratzner 1982; Kikuyama et al. 1988; Morstyn et al. 1983; Roberts et al. 1985), and cells which have taken it up can be stained using an immunoperoxidase technique after DNA denaturation. The percentage of stained cells can be counted to yield the 'S-phase labelling index'. However, the BUdR incubations are relatively complicated and time-consuming and the method is not easily carried out in a routine diagnostic service.

Markers of Quiescent and Non-proliferating Cells

An alternative approach to assessing the proliferative activity of cellular populations is to consider the fraction that is not participating in the cell cycle.

Entry into the cell cycle (i.e. G_0 to G_1 transition) is not only associated with the new or increased expression of genes, but it is also characterised by the reduced or complete abolition of expression or functional activity of other cellular genes or their products (Williams and Penman 1975). Some of these alterations in gene expression may represent key regulatory events in cell cycle control, while others could presumably represent structural alterations and other concomitant processes.

A variety of methods have been employed in attempts to identify gene products characteristic of non-proliferating cells. Immunisation of a mouse with the detergent insoluble extract of senescent human fibroblasts led to the production of a monoclonal antibody, S-30, that recognises a 57 kDa nuclear protein, designated statin, present only in non-proliferating cells (Wang 1985a). Immunoelectron and laser confocal microscopic studies have localised statin as a component of the nuclear envelope in senescent and quiescent cells (Wang 1985b; B. Ansari and P.A. Hall, personal communication). The expression of statin can be induced in young fibroblasts by culture to confluence with contact inhibition induced quiescence (Wang 1987) and the antigen is lost upon subculturing at low density or by serum-stimulated cell cycle entry, where the decline precedes transition from G_1 to S phase (Wang and Lin 1986). Statin has recently been reported to exist in two forms, based upon immunoprecipitation analysis of fractionated cellular extracts (Ching and Wang 1990). A detergent-soluble form of statin is present in both replicating and quiescent cultures of human fibroblasts, while a detergent-insoluble, nuclear-membrane associated form is detectable only in quiescent cultures. Serum stimulation of quiescent cells leads to a rapid decrease in detergent-insoluble statin associated with G_0 to S phase transition (Ching and Wang 1990).

Demonstration of statin immunoreactivity in cryostat sections of tissues reveals non-proliferating populations of cells (Wang and Krueger 1985), and the appearance of statin is associated with the process of differentiation (Bissonnette et al. 1990). It seems probable that the application of statin immunostaining may provide information regarding the existence of cellular subpopulations in a wide range of normal and pathological processes, and possibly aid in the definition of functional subpopulations in tumours. However, recent observations indicate that statin immunoreactivity may be present in cycling tumour cells, at least in vitro (see Chap. 3).

In addition, a range of genes expressed in growth-arrested cells has been defined by subtraction cDNA hybridisation methods (see Chap. 3; Ciccarelli et al. 1990; Manfioletti et al. 1990; Schneider et al. 1988). At present, little is known about the spatial and temporal distribution of the proteins encoded by these genes. Nuell et al. (1991) have described an intracellular 30 kDa protein (prohibitin) expressed in non-cycling cells, which appears to have regulatory properties and be widely conserved in evolution. Again, the details of the tissue distribution of prohibitin expression remain uncertain.

Not only should molecules such as prohibitin and statin be biologically interesting, they may also prove to be useful targets for the immunohistological characterisation of cellular subpopulations.

Markers of Transition from the Quiescent to the Proliferative State

A wide range of gene products have been identified which are associated with the transition from G_0 into G_1, including several cellular proto-oncogenes such as c-*fos*. A range of other quantitative and qualitative changes are associated with G_0 to G_1 transition: for example, Lau and Nathans (1985) identified by differential cDNA hybridisation a set of genes expressed during G_0 to G_1 transition, and the RNAs were found to accumulate and decay rapidly after stimulation of quiescent cells by serum growth factors. It is conceivable that immunohistological demonstration of gene products whose expression is tightly linked to the G_0 to G_1 transition (see Chaps 1 and 2) could be used as operational markers of cellular proliferation. However, none has so far been demonstrated to have any value in directly quantifying this process.

Quantification and Interpretation of Results

It is important to consider how to quantify results from these immunohisto-chemical staining techniques (Hall and Levison 1990; Hall and Woods 1990). Some authors have used manual counting of cells on tissue sections, and this can be rather time-consuming as a large number of cells must be counted to distinguish confidently between two values (Aherne et al. 1977). Sampling errors in counting cells can be minimised by using a random number table for selection of fields (Barnard et al. 1987), although the problems posed by tumour heterogeneity may merit examination of several representative blocks of tumour. Semi-quantitative methods have also been used, and these appear to give reproducible results in some studies (Garcia et al. 1987; Gatter et al. 1986; Jain et al. 1991; Shepherd et al. 1988). Image analysis techniques have also been applied to tissue sections and fine-needle aspirates immunostained with Ki-67 (Charpin et al. 1989; Franklin et al. 1987; Schwartz et al. 1989).

With any method of assessing cellular proliferation, the problem of heterogeneity within a specimen must be considered, particularly in tumours (Steel 1977). In addition, there are functionally and kinetically distinct populations in normal and pathological tissues (Hall and Watt 1989; Wright and Alison 1984). Although difficult to define (see Chap. 3), there is evidence that stem cells are present in tumours, and these may represent the predominant biologically and therapeutically relevant population (Buick and Pollack 1984; Mackillop et al. 1983). Consequently, the assessment of cell proliferation in different subpopulations may provide more meaningful information than counting all the cells in the sample as the denominator when deriving proliferation indices. A useful approach to the functional analysis of the proliferative characteristics of tissues may be to combine immunohistochemistry with double-staining methods for phenotypic markers. Heterogeneity may extend to there being different kinetic indices in tumour samples from different sites (McFarlane et al. 1986), and this may have considerable therapeutic implications (Norton 1985).

It must be recognised that, with the exception of the report by Galand and Degraef (1989), which employed sequential demonstration of tritiated thymidine incorporation and PCNA immunoreactivity, immunohistological methods only provide information about the "state" of proliferation (see Chap. 4, and Hall and Levison 1990). Information about "rate" of proliferation is considerably more difficult to obtain (Steel 1977; Wright and Alison 1984), but may be much more informative (see Chap. 7 for instance).

Conclusion

The different methods described in this chapter measure different things, and although comparable, the data produced are not identical. It is essential to consider the nature of the biological process being used as an index of proliferation. With enumeration of mitoses and with markers of DNA synthesis, this is relatively simple, but with many immunohistological markers, we remain uncertain as to exactly what is being measured. The clinical significance of different proliferative markers appears to vary in different types of neoplasm, probably reflecting the basic biological differences between tumour types (see Chap. 11). In non-Hodgkin's lymphomas, proliferative markers have, on the whole, been found to correlate well with histological grading (flow cytometry – Braylan et al. 1980; autoradiography – Silvestrini et al. 1977; transferrin receptor labelling – Habeshaw et al. 1983; Ki-67 immunostaining – Hall et al. 1988a; AgNORs – Hall et al. 1988b; PCNA immunostaining – Hall et al. 1990, Woods et al. 1991) and, where data were available, with clinical outcome (Costa et al. 1981; Hall et al. 1988a; Kvayloy et al. 1985; Roos et al. 1985; Woods et al. 1991). The various proliferative indices measured by different methods also seem to compare well with each other when performed in parallel on the same material (Hall et al. 1988b; Hall et al. 1990; Woods et al. 1991). Other tumours have not been so fully assessed with such a wide range of proliferative markers. However, there is some evidence that, in non-lymphoid neoplasms, the relationship of various proliferative markers with each other, and with prognosis, is not so straightforward. In gastric carcinoma, for example, a recent study has compared immunostaining for PCNA with DNA flow cytometry, and proliferative indices derived from these different methods show no correlation with each other (Jain et al. 1991). S-phase fraction and DNA ploidy ascertained by DNA flow cytometry do not show a significant correlation with clinical behaviour of the tumours. PCNA immunostaining, when assessed by manual counting of 1000 cells is also of limited value in determining prognosis, but the overall level of staining graded by a semi-quantitative method does appear to be a significant prognostic indicator. More extensive research is clearly needed to unravel the interrelationships between the various different methods of assessing cellular proliferation in different tumours, and their value in determining prognosis.

Although each of the individual techniques described has limitations, assessment of cell proliferation has numerous potential applications. As discussed above, there is considerable evidence that assessing cellular proliferation in tumours may provide information on prognosis, and multivariate analyses are urgently required to establish that proliferation is independent of other clinical and histological variables. The possibility that assessment of cellular proliferation during and after treatment may give clinically useful information about response should also be considered. Despite the numerous studies reporting the assessment of cell proliferation in tumours, such assessments do not yet directly influence patient management, with the possible exception of uterine and gastrointestinal smooth muscle tumours.

Another main area to which these techniques may be applied is in differentiating between lesions that fall into the "borderline" category between benign and malignant. Recent studies on gastric lesions have shown that the mean number of AgNORs per nucleus increases with the degree of cellular atypia from normal to dysplasia, and is highest in carcinoma (Rosa et al. 1990); immunohistochemical techniques should be evaluated in this and similar situations. A study of PC10 immunostaining in haemangiopericytomas, though involving small numbers of cases, has raised the possibility that this antibody may be of potential use in giving an indication of prognosis in "borderline" tumours of this type (Yu et al. 1991).

These methods of assessing proliferation are also of potential value in studying non-neoplastic tissue; they may provide information on the clinical outcome of certain pathological processes, e.g. disease activity in ulcerative colitis (Franklin et al. 1985) and recovery of renal function in acute tubular necrosis (B. Hartley et al. personal communication). These methods may also be of some help in differentiating between conditions that have very similar histological appearances: one such study has assessed immunostaining of renal transplant biopsies with Ki-67 and a monoclonal antibody to the transferrin receptor as a technique for distinguishing between cyclosporin toxicity and rejection (Seron et al. 1989).

The preservation of tissue morphology allows examination of cells for these proliferative markers in situ within histological sections and it is possible to correlate changes in proliferative activity with anatomical location, e.g. epithelial cells at different levels of the crypts in gastric mucosa (Brito et al. 1991). The recent advent of antibodies which can be used in conventionally fixed paraffin-embedded tissue opens the door to a limitless range of studies which are now possible on archival material available in all histopathology departments.

Acknowledgements. The authors would like to thank Mr Stefan Buk of the Department of Histopathology, Guy's Hospital, UMDS, for his technical expertise in producing the photomicrographs for this chapter.

References

Aherne WA, Camplejohn RS and Wright NA (1977) An introduction to cell population kinetics. Edward Arnold, London

Alberts B, Bray D, Lewis J, Raff M, Robert K and Watson JD (1989) Molecular biology of the cell, 2nd edn. Garland Publishing, New York

Baisch H and Gerdes J (1987) Simultaneous staining of exponentially growing versus plateau phase cells with the proliferation-associated antibody Ki-67 and propidium iodide: analysis by flow cytometry. Cell Tissue Kinet 20:387–391

Barnard NJ, Hall PA, Lemoine NR and Kadar NJ (1987) Proliferative index in breast carcinoma determined in situ by Ki-67 immunostaining and its relationship to pathological and clinical variables. J Pathol 152:287–295

Battersby S and Anderson TJ (1990) Correlation of proliferative activity in breast tissue using PCNA/cyclin [letter]. Hum Pathol 21:781

Bensch KG, Tanaka S, Hu S-Z, Wang TSF and Korn D (1982) Intracellular localisation of human DNA polymerase alpha with monoclonal antibodies. J Biol Chem 257:8391–8396

Bissonnette R, Lee M-J and Wang E (1990) The differentiation process of intestinal epithelial cells is associated with the appearance of statin, a non-proliferation-specific nuclear protein. J Cell Sci 95:247–255

Bravo R and Macdonald-Bravo H (1984) Induction of the nuclear protein cyclin in quiescent mouse 3T3 cells stimulated by serum and growth factors. Correlation with DNA synthesis. EMBO J 3:3177–3181

Bravo R and Macdonald-Bravo H (1985) Changes in the nuclear distribution of cyclin (PCNA) but not its synthesis depend on DNA replication. EMBO J 4:655–661

Bravo R and Macdonald-Bravo H (1987) Existence of two populations of cyclin/proliferating cell nuclear antigen during the cell cycle. Association with DNA replication sites. J Cell Biol 105:1549–1554

Bravo R, Frank R, Blundell PA and Macdonald-Bravo H (1987) Cyclin/PCNA is the auxiliary protein of DNA polymerase delta. Nature 326:515–517

Braylan RC, Diamond LW, Powell ML and Harty Golder B (1980) Percentage of cells in the S phase of the cell cycle in human lymphoma determined by flow cytometry. Correlation with labelling index and patient survival. Cytometry 1:171–174

Brito MJ, Filipe MI and Morris RW (1991) Bromodeoxyuridine (BrdU)-labelling in gastric carcinomas and non-involved gastric mucosa. J Pathol 163:174A

Brown DC and Gatter KC (1990) Monoclonal antibody Ki-67: its use in histopathology. Histopathology 17:489–503

Buick RN and Pollack MN (1984) Perspectives on clonogenic tumour cells, stem cells and oncogenes. Cancer Res 44:4909–4918

Celis JE and Celis A (1985) Individual nuclei in polykaryons can control cyclin distribution and DNA synthesis. EMBO J 4:1187–1192

Celis JE and Madson P (1986) Increased nuclear PCNA/cyclin antigen staining of non-S phase transformed human amnion cells engaged in nucleotide excision DNA repair. FEBS Lett 209: 277–283

Chang CD, Ottavio L, Travalli S, Lipson KE and Baserga R (1990) Transcriptional and post-transcriptional regulation of the proliferating cell nuclear antigen gene. Mol Cell Biol 10:3289–3296

Charpin C, Andrac L, Habib M-C et al. (1989) Immunodetection in fine-needle aspirates and multiparametric (SAMBA) image analysis. Receptors (monoclonal antiestrogen and anti-progesterone) and growth fraction (monoclonal Ki-67) evaluation in breast carcinomas. Cancer 63:863–872

Ching G and Wang E (1990) Characterisation of two populations of statin and the relationship of their synthesis to the steady state of cell proliferation. J Cell Biol 110:255–261

Chou MY, Chang ALC, McBride J, Donoff B, Gallagher GT and Wong DTW (1990) A rapid method to determine proliferation patterns of normal and malignant tissues by H^3 mRNA in situ hybridisation. Am J Pathol 136:729–733

Ciccarrelli C, Philipson L and Sorrentino V (1990) Regulation of growth arrest-specific genes in mouse fibroblasts. Mol Cell Biol 10:1525–1529

Costa A, Bonadonna G, Villa E, Valagussa P and Silverstrini R (1981) Labelling index as a prognostic marker in non-Hodgkin's lymphoma. J Natl Cancer Inst 66:1–5

Fairman MP (1990) DNA polymerase delta/PCNA: actions and interactions. J Cell Sci 95:1–4

Falini B, Canino S, Sacchi S et al. (1988) Immunocytochemical evaluation of the percentage of proliferating cells in pathological bone marrow and peripheral blood samples with the Ki-67 and antibromo-deoxyuridine antibodies. Br J Haematol 69:311–320

Franklin WA, McDonald GB, Stein HO et al. (1985) Immunohistologic demonstration of abnormal colonic cell kinetics in ulcerative colitis. Hum Pathol 16:1129–1132

Franklin WA, Bibbo M, Doria MI et al. (1987) Quantitation of estrogen receptor and Ki-67 staining in breast carcinoma by the microTICAS Image analysis system. Anal Quant Cytol Histol 9:279–286

Galand P and Degraef C (1989) Cyclin/PCNA immunostaining as an alternative to tritiated thymidine pulse labelling for marking S phase cells in paraffin sections from animal and human tissues. Cell Tissue Kinet 22:383–392

Garcia CF, Weiss LM, Lowder J et al. (1987) Quantitation and estimation of lymphocyte subsets in tissue sections. Comparison with flow cytometry. Am J Clin Pathol 87:470–477

Gatter KC, Brown G, Trowbridge IS, Woolston RE and Mason DY (1983) Transferrin receptors in human tissues: their distribution and possible clinical significance. J Clin Pathol 36:539–545

Gatter KC, Dunnill MA, Gerdes J, Stein H and Mason DY (1986) New approach to assessing lung tumours in man. J Clin Pathol 39:590–593

Gerdes J, Schwab U, Lemke H and Stein H (1983) Production of a monoclonal antibody reactive with a human nuclear antigen associated with cell proliferation. Int J Cancer 31:13–20

Gerdes J, Lemke H, Wacker HH, Schwab J and Stein H (1984) Cell cycle analysis of a cell proliferation associated human nuclear antigen defined by the monoclonal antibody Ki-67. J Immunol 133:1710–1715

Gratzner HG (1982) Monoclonal antibody to 5-bromo and 5-iodo-deoxyuridine; a new reagent for detection of DNA replication. Science 218:474–475

Habeshaw JA, Lister TA, Stansfeld AG and Greaves MF (1983) Correlation of transferrin receptor expression with histological class and outcome in non-Hodgkin's lymphoma. Lancet i(8323):498–501

Hadjiolova KV, Martinova YS, Yankulov KY, Davidov V, Kancheva LS and Hadjiolov AA (1989) An immunocytochemical study of proliferating nuclear matrix antigen p125/6.6 during rat spermatogenesis. J Cell Sci 93:173–177

Hall PA and Levison DA (1990) Assessment of cell proliferation in histological material. J Clin Pathol 43:184–192

Hall PA and Watt FM (1989) Stem cells: the generation and maintenance of cellular diversity. Development 106:619–633

Hall PA and Woods AL (1990) Immunohistological markers of cell proliferation. Cell Tissue Kinet 23:531–549

Hall PA, Richards MA, Gregory WM, d'Ardenne AJ, Lister TA and Stansfield AG (1988a) The prognostic value of Ki-67 immunostaining in non-Hodgkin's lymphoma. J Pathol 154:223–235

Hall PA, Crocker J, Watts A and Stansfeld AG (1988b) A comparison of nucleolar organiser region staining and Ki-67 immunostaining in non-Hodgkin's lymphoma. Histopathology 12: 373–381

Hall PA, Levison DA, Woods AL et al. (1990) Proliferating cell nuclear antigen (PCNA) immuno-localisation in paraffin sections: an index of cell proliferation with evidence of de-regulated expression in some neoplasms. J Pathol 162:285–294

Jain S, Filipe MI, Hall PA, Waseem NH, Lane DP and Levison DA (1991) Proliferating cell nuclear antigen (PCNA) prognostic value in gastric carcinoma. J Clin Pathol 44:655–659

Jaskulski D, Gatti G, Travali S, Calabretta B and Baserga R (1988) Regulation of the proliferating cell nuclear antigen cyclin and thymidine kinase mRNA levels by growth factors. J Biol Chem 263:10175–10179

Kamel OW, Franklin WA, Ringus JC and Meyer JS (1989) Thymidine labelling index and Ki-67 growth fraction in lesions of the breast. Am J Pathol 134:107–113

Kikuyama S, Kubora T, Watanabe M, Ishibiki K and Abe O (1988) Cell kinetic study of human carcinomas using bromodeoxyuridine. Cell Tissue Kinet 20:1–6

Kvaloy S, Morton PF, Kaalhum O, Hoie J, Foss-Abrahamsen A and Godal T (1985) Spontaneous (³H)-thymidine uptake in B cell lymphomas. Relationship to treatment response and survival. Scand J Haematol 34:429–435

Lau LF and Nathans D (1985) Identification of a set of genes expressed during the G_0/G_1 transition of cultured mouse cells. EMBO J 4:3145–3151

Lee MYWT, Alejandro R and Toomey NL (1989) Immunochemical studies of DNA polymerase delta: relationships with DNA polymerase alpha. Arch Biochem Biophys 272:1–9

Lloyd RV, Wilson BS, Varani J, Gaur PK, Moline S and Makari JG (1985) Immunocytochemical characterisation of a monoclonal antibody that recognises mitosing cells. Am J Pathol 121:275–283

Mackillop WJ, Ciampi A, Till JE and Buick RN (1983) A stem cell model of human tumour growth: implications for tumour cell clonogenic assays. J Natl Cancer Inst 70:9–16

Manfioletti G, Ruaro ME, Del Sal G, Philipson L and Schneider C (1990) A growth arrest-specific gene (gas)codes for a membrane protein. Mol Cell Biol 10:2924–2930

Marraccino RL, Wahl AF, Keng PC, Lord EEM and Bambara RA (1987) Cell cycle dependent activation of polymerase alpha and delta in Chinese hamster ovary cells. Biochemistry 26:7864–7870

McFarlane JH, Quirke P and Bird CC (1986) Flow cytometric analysis of DNA heterogeneity in non-Hodgkin's lymphoma. J Pathol 149 (abstr):236

Miyachi K, Fritzler MJ and Tan EM (1978) Autoantibody to a nuclear antigen in proliferating cells. J Immunol 121:2228–2234

Morris GF and Mathews MB (1989) Regulation of proliferating cell nuclear antigen during the cell cycle. J Cell Biol 264:13856–13864

Morstyn G, Hsu SM, Kinsella T, Gratzner A, Russo A and Mitchell JB (1983) Bromodeoxyuridine in tumours and chromosomes detected with a monoclonal antibody. J Clin Invest 72:1844–1985

Mushika M, Shibata K, Miwa T, Suzuoki Y and Kaneda T (1989) Proliferative cell index in endometrial adenocarcinoma of different nuclear grades. Jpn J Cancer Res 80:223–227

Namikawa R, Suchi T, Ueda R, Itoh G, Ota K and Takahashi T (1987a) Phenotyping of proliferating lymphocytes in angioimmunoblastic lymphadenopathy and related lesions by the double immunoenzymatic staining technique. Am J Pathol 127:279–287

Namikawa R, Ueda R, Suchi T, Itoh G, Ota K and Takahashi T (1987b) Double immunoenzymatic detection of surface phenotype of proliferating lymphocytes in situ with monoclonal antibodies against DNA polymerase alpha and lymphocyte membrane antigens. Am J Clin Pathol 87:725–731

Nishida C, Reinhard P and Linn S (1988) DNA repair synthesis in human fibroblasts requires DNA polymerase delta. J Biol Chem 263:501–510

Norton L (1985) Implications of kinetic heterogeneity in clinical oncology. Semin Oncol 12:231–249

Nuell MJ, Stewart DA, Walker L et al. (1991) Prohibitin, an evolutionarily conserved intracellular protein that blocks DNA synthesis in normal fibroblasts and HeLa cells. Mol Cell Biol 11:1372–1381

Ogata K, Kurki P, Celis JE, Nakamura RM and Tan EM (1987) Monoclonal antibodies to a nuclear protein (PCNA/cyclin) associated with DNA replication. Exp Cell Res 168:475–486

Ottavio L, Chang CD, Rizzo MG, Travalli S and Casadevall C (1990) Importance of introns in the growth regulation of mRNA levels of the proliferating cell nuclear antigen gene. Mol Cell Biol 10:303–309

Philipova RN, Zhelev NZ, Todorov IT and Hadjiolov AA (1987) Monoclonal antibodies against nuclear matrix antigen in proliferating human cells. Biol Cell 60:1–8

Pileri S, Gerdes J and Rivano M (1987) Immunohistochemical determination of growth fractions in human permanent cell lines and lymphoid tumours: a critical comparison of the monoclonal antibodies OKT9 and Ki-67. Br J Haematol 68:271–276

Prelich G, Tan C-K, Kostura M et al. (1987) Functional identity of proliferating cell nuclear antigen and DNA polymerase delta auxiliary protein. Nature 326:517–520

Quak JJ, Van Dongen G, Koken MAE et al. (1990) Identification of a 43-kDa nuclear antigen associated with proliferation by monoclonal antibody K 112. Int J Cancer 46:50–55

Roberts CG, deFazio A and Tattersall MH (1985) A simple and rapid method for the identification of cycling cells in freshly excised tumours. Cytobios 43:313–318

Rosa J, Mehta A and Filipe MI (1990) Nucleolar organizer regions in gastric carcinoma and its precursor stages. Histopathology 16:265–269

Roos G, Dige U, Lenner P, Lindh J and Johansson H (1985) Prognostic significance of DNA analysis by flow cytometry in non-Hodgkin's lymphoma. Haematol Oncol 3:233–242

Sasaki K, Murakami R, Kawasaki M and Takahashi M (1987) The cell cycle associated change of the Ki-67 reactive nuclear antigen expression. J Cell Physiol 133:579–584

Sasaki K, Matsumura K, Tsuji T, Shinozaki F and Takahashi M (1988) Relationship between labelling indices of Ki-67 and BrdUrd in human malignant tumours. Cancer 62:989–993

Schneider C, King RM and Philipson L (1988) Genes specifically expressed at growth arrest in mammalian cells. Cell 54:787–793

Schrape S, Jones DB and Wright DH (1987) A comparison of three methods for the determination of the growth fraction in non-Hodgkin's lymphomas. Br J Cancer 55:283–286

Schwarting R, Gerdes J, Niehus J, Jaesche L and Stein H (1986) Determination of the growth fraction in cell suspensions by flow cytometry using the monoclonal antibody Ki-67. J Immunol Methods 90:365–371

Schwartz BR, Pinkus G, Bacus S, Toder M and Weinberg DS (1989) Cell proliferation in non-Hodgkin's lymphomas. Digital image analysis of Ki-67 antibody staining. Am J Pathol 134:327–336

Scott RJ, Hall PA, Haldane J et al. (1991) A comparison of immunohistochemical markers of cell proliferation with experimentally determined growth fraction. J Pathol (in press)

Seron D, Alexopoulos E, Raftery MJ, Hartley RB and Cameron JS (1989) Diagnosis of rejection in renal allograft biopsies using the presence of activated and proliferating cells. Transplantation 47:811–816

Shepherd NA, Richman PI and England J (1988) Ki-67 derived proliferative index in colorectal adenocarcinoma with prognostic correlations. J Pathol 155:213–219

Silvestrini R, Piazza R, Riccardi A and Rilke F (1977) Correlation of kinetic findings with morphology of non-Hodgkin's lymphomas. J Natl Cancer Inst 58:499–504

Silvestrini R, Costa A, Vereroni S, Del Bino G and Persici P (1988) Comparative analysis of different approaches to investigate cell kinetics. Cell Tissue Res 21:123–131

Steel GG (1977) Growth kinetics of tumours. Clarendon Press, Oxford

Suzuka I, Daidoji H, Matsuoka M et al. (1989) Gene for proliferating cell nuclear antigen (DNA polymerase-delta auxiliary protein) is present in both mammalian and higher plant genomes. Proc Natl Acad Sci (USA) 96:3189–3193

Syraoja J and Linn S (1989) Characterisation of a large form of DNA polymerase delta from HeLa cells that is insensitive to PCNA. J Biol Chem 264:2489–2497

Takasaki Y, Deng J-S and Tan EM (1987) A nuclear antigen associated with cell proliferation and blast transformation. Its distribution in synchronised cells. J Exp Med 154:1899–1909

Tan CK, Castillo C, So AG and Downey KM (1986) An auxiliary protein for DNA polymerase delta from fetal calf thymus. J Biol Chem 261:12310–12316

Tanaka S, Hu S-Z, Wang TSF and Korn D (1982) Preparation and preliminary characterisation of monoclonal antibodies against DNA polymerase alpha. J Biol Chem 257:8386–8390

Todorov IT, Philipova RN, Zhelev NZ and Hadjiolov AA (1988) Changes in a nuclear matrix antigen during the cell cycle: interphase and mitotic cells. Biol Cell 62:105–110

Toschi L and Bravo R (1988) Changes in cyclin/PCNA distribution during DNA repair synthesis. J Cell Biol 107:1623–1628

Verheijen R, Kuijpers HJH, Schlingeman RO et al. (1989a) Ki-67 detects a nuclear matrix associated proliferation antigen. I. Intracellular localisation during interphase. J Cell Sci 92:123–130

Verheijen R, Kuijpers HJH, Van Driel R et al. (1989b) Ki-67 detects a nuclear matrix associated proliferation antigen. II. Localisation in mitotic cells and association with chromosomes. J Cell Sci 92:531–540

Veroni S, Costa A, Motta R, Giardini R, Jilke F and Silvestrini R (1988) Comparative analysis of ^{3}H thymidine labelling index and monoclonal antibody Ki-67 in non-Hodgkin's lymphoma. Haematol Oncol 6:21–28

Wahl AF, Geis AM, Spain BH, Wong SW, Korn D and Wang TSF (1989) Gene expression of human DNA polymerase alpha during cell proliferation and the cell cycle. Mol Cell Biol 8:5016–5025

Walker RA and Camplejohn RS (1988) Comparison of monoclonal antibody Ki-67 reactivity with grade and DNA flow cytometry of breast carcinomas. Br J Cancer 57:281–283

Wang E (1985a) A 57000 mol-Wt protein uniquely present in nonproliferating cells and senescent human fibroblasts. J Cell Biol 100:545–551

Wang E (1985b) Rapid disappearance of statin, a non-proliferating and senescent cell-specific protein, upon reentering the process of cell cycling. J Cell Biol 101:1695–1701

Wang E (1987) Contact inhibition-induced quiescent state is marked by intense nuclear expression of statin. J Cell Physiol 133:151–157

Wang E and Krueger JG (1985) Application of a unique monoclonal antibody as a marker for non-proliferating subpopulations of cells of some tissues. J Histochem Cytochem 33:587–594

Wang E and Lin SL (1986) Disappearance of statin, a protein marker for non-proliferating cells and senescent cells, following serum-stimulated cell cycle entry. Exp Cell Res 167:135–143

Waseem NH and Lane DP (1990) Monoclonal antibody analysis of the proliferating cell nuclear antigen (PCNA). Structural conservation and the detection of a nucleolar form. J Cell Sci 96:121–129

Wersto RP, Herz F, Gallagher RE and Koss LG (1988) Cell cycle dependent reactivity with the monoclonal antibody Ki-67 during myeloid differentiation. Exp Cell Res 179:79–88

Williams JG and Penman S (1975) The messenger RNA sequences in growing and resting mouse fibroblasts. Cell 6:197–206

Woods AL, Hall PA, Shepherd NA, Hanby AM, Waseem NH, Lane DP and Levison DA (1991) The assessment of proliferating cell nuclear antigen (PCNA) immunostaining in primary gastro-intestinal lymphomas and its relationship to histological grade, S + G$_2$M phase fraction (flow cytometric analysis) and prognosis. Histopathology 19:21–27

Wright NA and Alison M (1984) The biology of epithelial cell populations. Clarendon Press, Oxford

Yamaguchi A, Takegawa S, Ishida T et al. (1990) Detection of the growth fraction in colorectal tumours by a monoclonal antibody against DNA polymerase alpha. Br J Cancer 61:390–393

Yankulov KY, Hadjiolov KV, Kounev KV, Getov CHR and Hadjiolov AA (1989) The use of monoclonal antibody against a proliferating cell nuclear matrix antigen in the study of solid human tumours. Int J Cancer 43:800–804

Yu CC-W, Hall PA, Fletcher CDM et al. (1991) Haemangiopericytomas. The prognostic value of immunohistochemical staining with a monoclonal antibody to proliferating cell nuclear antigen (PCNA). Histopathology 19:29–33

9 Nucleolar Organiser Regions

J.C.E. UNDERWOOD

Introduction

Nucleolar organiser regions (NORs) are the genomic DNA segments encoding for ribosomal RNA. They can be visualised in chromosome preparations and in interphase nuclei because each NOR is associated with argyrophilic protein; the silver-stained structures thus demonstrated are called AgNORs.

Recent work has suggested that AgNOR counts in interphase nuclei may be of value in histopathology and cytopathology for three interrelated purposes:

1. To assist in the distinction between malignant and benign or reactive conditions
2. As a prognostic index in malignant neoplasms
3. As a measure of proliferative state

Although abnormal ploidy can result in a variation in the number of NORs within a nucleus (as in trisomy-21), for reasons given below the consensus view is that variations in AgNOR counts in interphase nuclei are more likely to be influenced by the proliferative state of the cell.

The purpose of this chapter is to outline briefly the methodology for demonstrating and counting AgNORs and to summarise their utility in histopathology for estimating the cellular proliferative state.

Nucleolar Organiser Regions and Nucleoli

A eukaryotic cell typically produces approximately 10^6 copies of each type of rRNA. Since rRNA rather than protein is the end-product (protein synthesis can be amplified at the ribosomal level), this abundance of rRNA can be accomplished only by the transcription of multiple copies of the rRNA genes. Human cells contain, therefore, approximately 200 rRNA gene copies per haploid genome; these are clustered in tandem, separated by untranscribed spacer DNA, on five chromosomes. Each rRNA gene cluster is known as a

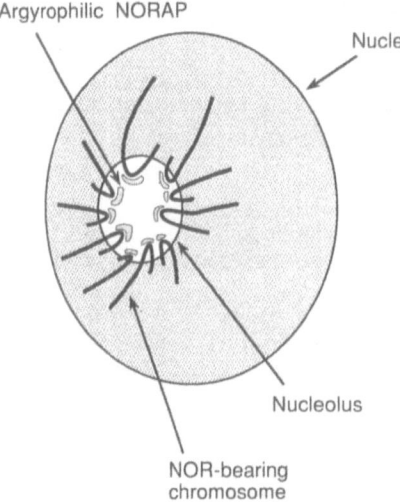

Figure 9.1. Simplified diagrammatic representation of a nucleolus to show the relationship between NORs and the silver-binding nucleolar organiser region-associated proteins. The NORs are located on the acrocentric chromosomes which project as loops into the nucleolus. Transcriptionally-active NORs are each characterised by an aggregate of NOR-associated proteins (NORAPs) some of which are argyrophilic and, therefore, designated AgNORs.

nucleolar organiser region (NOR). In the human karyotype, the NORs are located on the five acrocentric chromosomes – 13, 14, 15, 21 and 22 (Alberts et al. 1989).

In an interphase nucleus, the NORs form loops of DNA projecting into one or more nucleoli (Fig. 9.1); the NORs correspond to the fibrillar centres visible by electron microscopy. RNA transcription occurs in the dense fibrillar component surrounding the fibrillar centre of the nucleolus. The granular component of the nucleolus harbours the maturing ribosomes.

The rRNA genes are transcribed by RNA polymerase I, one of several NORAPs. Another NORAP is nucleolin or C_{23} protein; this is the principal argyrophilic NORAP. Nucleolin is an RNA-binding protein, the amount of which corresponds to the level of transcriptional activity. However, nucleolin can remain in the nucleus for some time after transcription has ceased.

Argyrophilic NORs

The argyrophilia of the nucleolus was first discovered in the late nineteenth century (Ruzicka 1891, cited by Busch et al. 1979). With prolonged silver staining the entire nucleolus can be rendered black on microscopy with trans-mitted light, but this obscures the individual argyrophilic NORs that should be visible as little black dots within each nucleolus. The silver-stained black dots thus demonstrated are referred to as AgNORs (Fig. 9.2). The argyrophilia of NORs is attributable principally to the affinity of nucleolin for silver.

The AgNOR technique was used extensively by cytogeneticists before it was applied to histopathological material. Goodpasture and Bloom (1975) and Howell (1977) were largely responsible for the cytogenetic applications of the argyrophilic NOR reaction to chromosome preparations.

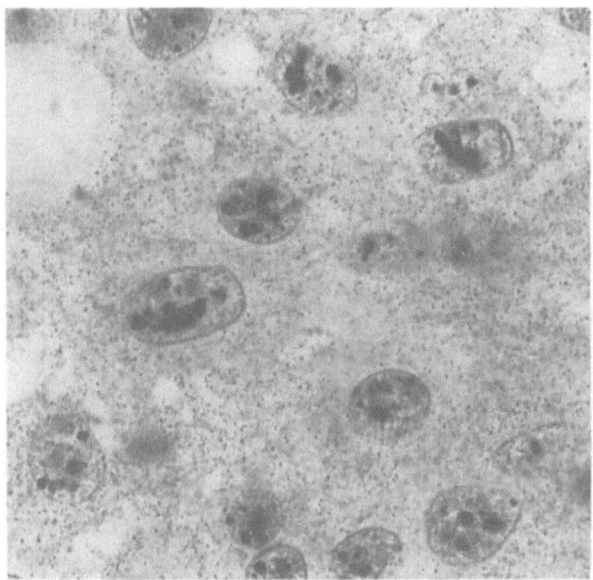

Figure 9.2. AgNOR dots in interphase nuclei of an invasive carcinoma of the breast.

The potential of the AgNOR reaction for examining NORs in interphase nuclei was beautifully illustrated by Busch et al. (1979) who, using a two-step silver-staining technique, clearly demonstrated numerous tiny black dots within the nuclei of air-dried Carnoy's-fixed tumour cell lines. The widespread application of the AgNOR technique to histopathology was facilitated by the development of a simplified one-step silver-staining method (Ploton et al. 1986).

Demonstration of NORs

NORs can be demonstrated by a variety of methods. The earliest method was in situ hybridisation using radiolabelled rRNA to detect the complementary rDNA sequences in NORs; more recently, non-isotopically labelled rRNA has been used, giving greater resolution.

Antibodies to NORAPs have been developed for the immunohistochemical localisation of NORs, but currently this strategy has little advantage over the simpler AgNOR technique and is not used widely.

It is important to emphasise that any technique aimed at visualising NORAPs in interphase nuclei, such as immunohistochemistry or the AgNOR technique, may underestimate the actual number of NOR sites in the genome. This is because only actively transcribing NORs are associated with sufficient specific protein to be visible by these methods. Only probing with sense-sequence rRNA will reveal the actual number of NORs, because the hybridisation with rDNA is independent of transcriptional activity.

The most widely used method for investigating NORs in interphase nuclei is the silver-staining technique for AgNORs.

AgNOR Staining Techniques

Goodpasture and Bloom (1975) and Howell (1977) are credited with the first descriptions of the AgNOR technique; this was developed initially for cyto-genetic studies on chromosome spreads. During the following decade the method was simplified and adapted by Ploton et al. (1986) to demonstrate AgNORs in interphase nuclei in tissue sections. Latterly, the AgNOR tech-nique has been promulgated by Crocker and his colleagues (Crocker 1990) for use in histopathology, spawning numerous publications.

The simplest one-stage staining method involves treating de-waxed rehy-drated tissue sections with a silver nitrate solution of the following composi-tion: 2 volumes of aqueous silver nitrate (50 g/dl) with 1 volume of aqueous formic acid (1 g/dl) containing gelatin at 2 g/dl. Sections are exposed to this solution for 30–60 min at room temperature in the dark or under safelight conditions. After thorough washing the sections are dehydrated, cleared and mounted in the usual way. A counterstain is optional. The preparations appear to be permanent.

Two modifications have been proposed to reduce non-specific deposition of silver grains on the section. First, the staining reaction can be performed with the section inverted; any loosely bound silver grains fall away (Coghill et al. 1990). Second, a celloidin film can be used as a permeable barrier between the section and the staining solution, preventing the non-specific accumulation of silver grains (Chiu et al. 1989).

The most critical aspect of the AgNOR technique is probably the staining time. If this is excessive, the minute individually discernible black dots may coalesce; the AgNORs within a nucleolus can be easily obscured in this way. The optimum staining time is best determined separately for each tissue and fixation method.

Formalin seems to be an adequate fixative for AgNOR staining and most studies have been performed on archival tissue that has been fixed in this way. Stronger staining may be seen in sections of tissue that have been fixed in al-cohol fixatives, but it is claimed that the number of AgNOR dots is unaffected (Smith et al. 1988). However, Griffiths et al. (1989) have criticised the use of formalin-fixed tissue for AgNOR studies because they found evidence of coalescence of smaller particles and poor correlation between the number of AgNOR dots and prognosis, cell proliferation or DNA ploidy.

Application to Cytology and Tissue Imprints

The method, with appropriate modification, gives particularly clean results with cytological preparations. Cytological preparations guarantee also the presence of entire nuclei for AgNOR enumeration. Furthermore, the flattening of the nuclei tends to spread out the AgNORs slightly so that they can be counted individually more easily. Comparison of imprints and sections of non-Hodgkin's lymphomas reveals higher AgNOR counts in the former – 16.3 per nucleus in imprints, 6.0 per nucleus in sections (Boldy et al. 1989).

Enumeration of AgNORs

AgNORs appear as black dots within nuclei. They are small, usually less than 1 μm in diameter individually, but larger aggregates may be seen. There is debate about the best method for counting AgNOR dots, but first it is important to consider the effects of sampling, section thickness, NOR aggregation and segregation, ploidy, and transcriptional activity on the number and distribution of AgNORs within nuclei.

Tissue Sampling

As with all proliferation indices, there are likely to be regional variations within a tumour (see Chap. 3). The tumour periphery is likely to be the most proliferative zone, simply because it will be relatively well nourished. AgNOR counts may well be higher at the tumour periphery than at or near the centre (Quinn and Wright 1990). When comparing lesions the choice of zone should be standardised.

Section Thickness

In theory it might be better to use very thin sections and thereby reduce the risk of obscuring some AgNORs by superimposition. However, this would result in an underestimate of the total AgNORs per nucleus because only part of each nucleus would be present within the section. This tendency to underestimate would be exaggerated in malignant neoplasms because these typically have larger nuclei than do benign lesions; thin sections would contain relatively smaller samples of malignant nuclei.

Using thicker sections (3–5 μm) would avoid the risk of underestimating the AgNOR number per nucleus due to the previous effect (nuclear sample size), but it still may result in an underestimate due to superimposition of AgNORs with the nucleus.

NOR Aggregation and Segregation

During the mitotic cycle there is aggregation and segregation of NORs and, consequently, AgNORs (Fig. 9.3). Immediately after mitosis the NORs are dispersed through the nucleus and the nucleolus is not readily apparent; at this stage, AgNOR staining would reveal a relatively large number of dots. The NORs then cluster to form one or more nucleoli; AgNOR staining reveals fewer dots because the NORs have coalesced. In late G_2, the NORs tend to disperse with dissolution of the nucleolus. The most extreme segregation of NORs is seen, of course, during mitosis when the chromosomes separate.

The apparent number of NORs within a nucleus may, therefore, vary because of changes in the ease with which they can be counted when aggregated or segregated; the AgNOR count may be higher in cells in late G_2 or early G_1

Phase of
cell cycle

G1 ⟶ S ⟶ G2 ⟶ M ⟶ G1

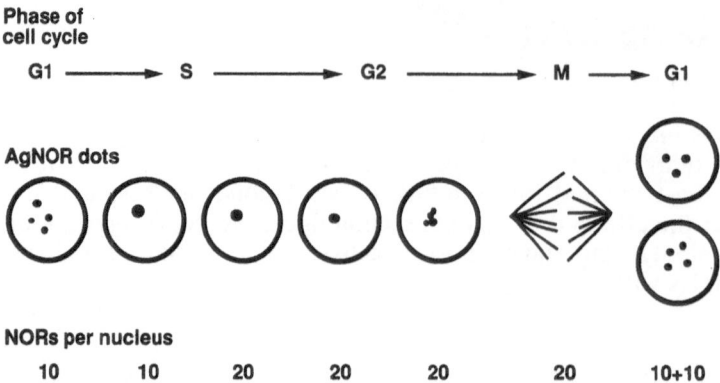

AgNOR dots

NORs per nucleus

| 10 | 10 | 20 | 20 | 20 | 20 | 10+10 |

Figure 9.3. AgNOR aggregation and segregation through the cell cycle and in relation to the actual number of NORs present. Note that the number of discernible AgNOR dots does not necessarily correspond to the actual number of NOR sites in the nucleus.

simply because the NORs are segregated and more likely to be individually discernible than at other stages in the cell cycle.

NOR Numbers and Ploidy

During the S phase of the cell cycle there is chromosomal replication as a prelude to mitotic division. Along with all other chromosomes this leads to a temporary doubling of the number of acrocentric chromosomes and, therefore, a doubling of the NOR number per nucleus. In theory, this might lead to an abrupt increase in the AgNOR count between G_1 and G_2 cells. However, because the NORs tend naturally to aggregate within nucleoli, in *normal* cells this does not appear to be significantly responsible for fluctuations in the AgNOR count in interphase nuclei.

In *neoplastic* nuclei the cyclical doubling of ploidy may be one factor explaining the higher AgNOR counts observed. This could be due to a concurrent defect of nucleolar association, causing the newly-formed NORs to remain dispersed; this is, however, speculation.

Abnormal ploidy can result in an elevated AgNOR count only if it involves multiplication of the NOR-bearing acrocentric chromosomes or the appearance of novel NOR sites on other chromosomes.

Transcriptional Activity

If in situ hybridisation is used to demonstrate NORs in interphase nuclei, then transcriptional activity has no influence on the number that are visible. However, the most popular method for demonstrating NORs relies not on the presence of the NORs themselves, but on the argyrophilic NOR-associated proteins, the amount of which will be influenced by the level of transcriptional activity. Thus, in *normal* cells there is a discrepancy: the AgNOR count tends to underestimate the actual number of NOR sites in the nucleus, particularly in late G_2 when the transcriptional activity of rRNA and other genes is likely to be declining prior to mitosis. In *neoplastic* cells transcriptional activity may be sustained at a high level through the cell cycle.

Assessment of AgNORs in Interphase Nuclei

The commonest AgNOR indices are mean number and area per nucleus. Before examining the way in which these indices are derived, it is imperative to consider the influence of sample size on precision of the AgNOR measurements and the effect of superimposition and aggregation on the interpretation of AgNOR-stained preparations.

Sample Size

As with all quantitative procedures, sample size is a major factor in determining the precision of the final estimate. In some AgNOR studies, the number of nuclei assessed seems to have been arbitrarily chosen as 100 or 200. The simplest way of ensuring that a mean estimated AgNOR count has acceptable precision ($\pm 5\%$) is to perform a cumulative means procedure. It may be necessary to assess thousands of nuclei to reveal small differences between groups of lesions at this acceptable level of significance.

Finally, it is important to ensure that the nuclei being assessed for AgNOR status within a tumour are indeed tumour cell nuclei and that the AgNOR count is not contaminated with data from observations on infiltrating inflammatory cells or stromal elements.

Effect of AgNOR Superimposition and Aggregation

AgNORs are often aggregated within one or more nucleoli or they may appear to be dispersed randomly through the nucleus. When they are tightly aggregated within nuclei, they are extremely difficult to count individually because of superimposition or actual coalescence. For this reason, many investigators make no attempt to count the *actual* number of AgNORs because this procedure has an unacceptable level of fatigue and inter-observer error. The alternative is to count each individually discernible AgNOR dot, each of which may be a single AgNOR or a cluster of AgNORs; the AgNOR cluster within a nucleolus has been abbreviated to "AgNu" (silver-stained nucleolus). This counting strategy inevitably underestimates the number of NORs per nucleus, but it may be a more reliable method of AgNOR enumeration for diagnostic and prognostic purposes.

AgNOR counting is usually easier with cytological preparations because there is less non-specific silver deposition and the dots are more dispersed.

AgNOR Number per Nucleus

The most widely derived AgNOR index is the mean number per nucleus. In normal tissues and benign neoplasms, this number is usually less than that obtained from malignant neoplasms. Within a group of malignant neoplasms, the mean AgNOR number per nucleus often correlates with indices of proliferative activity and with prognosis. Although, for reasons already given, the AgNOR count does not correspond precisely to actual number of NOR sites within the cell, this does not invalidate the technique when it is used to compare lesions stained and assessed in the same way.

Faced with the practical problem of counting AgNOR dots within a nucleus, there are two main options (Fig. 9.4):

1. To count each dot as one and sum to derive a total
2. To count all the intranucleolar dots together as one (i.e. one AgNu), each extranucleolar dot as one, and sum to derive a total

In theory, the first option is likely to be more discriminating because it attempts to quantify the actual number of AgNORs within each nucleus. However, it suffers from the disadvantages of being more time-consuming and associated with a higher degree of inter-observer error. This is because the individual AgNORs aggregated within nucleoli are difficult to resolve separately.

The second option is associated with less inter-observer error, but it underestimates the actual AgNOR content of the nucleus. This option is probably more a measure of AgNOR dispersion than absolute AgNOR number.

A standardised method for AgNOR counting has been proposed (Crocker et al. 1989). In summary, it is recommended that individual dots – even those within nucleoli – are counted, since experience has shown that this is likely to be more discriminating between benign and malignant lesions.

AgNOR Area

AgNOR area may be an equally legitimate index. Total nuclear AgNOR area will be influenced by two factors – the number of AgNOR dots in the nuclear profile and their diameter. The diameter of AgNOR dots is almost certainly

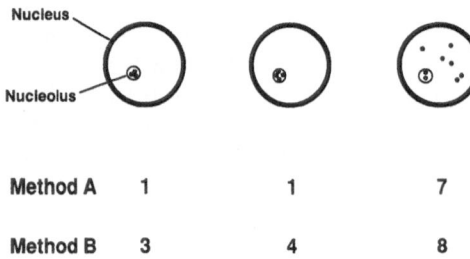

Method A	1	1	7
Method B	3	4	8

Figure 9.4. Alternative methods for counting AgNOR dots. *Method A*: because of the difficulty of counting individually each intranucleolar dot, those aggregated within each nucleus are summed as one, AgNOR dots outside the nucleoli are each counted as one, and a total is derived. *Method B*: All AgNOR dots are counted, each as one.

increased by the level of transcriptional activity causing the accumulation of larger pools of argyrophilic NORAPs, and reduced by the dispersion of individual AgNORs within large confluent argyrophilic structures.

Quantification by interactive image analysis of AgNORs in photomicrographs of non-Hodgkin's lymphomas enabled Crocker and Egan (1988) to examine the relationship between individual AgNOR area and number. In low-grade lesions the mean AgNOR area ranged from 0.48 to $1.99\,\mu m^2$, and in high-grade lesions from 0.33 to $0.51\,\mu m^2$. There was a strong inverse relationship between AgNOR area and number per nucleus. The combination of AgNOR area and number offers a powerful discriminant between high- and low-grade non-Hodgkin's lymphomas.

AgNOR Assessment by Image Analysis

The high optical density of the silver-stained dots is ideal for measurement of area by image analysis. However the small size of AgNOR dots means that many image analysis systems will be working at or close to the limit of resolution determined by their pixel size. This limit can be compensated for, to some extent, by using enlarged photomicrographs. Another problem with image analysis is the necessity to find the optimum plane of focus. Not all AgNOR dots will be in the same plane of focus, and human judgement may be necessary to select appropriate cells and planes for measurement and then "clean up" the image by editing.

Rüschoff et al. (1990) used a Cue 2 automatic image analyser (Olympus Optical Co., Hamburg) and a final screen magnification of ×4000. The AgNOR dots were then discriminated from the background by grey-value thresholding. Reproducibility was tested by the cumulative means procedure. Accuracy to within 5% could be achieved for AgNOR number and area by assessing approximately only 50 nuclei.

Significance of High AgNOR Counts

Because the *AgNOR count*, i.e. mean number of AgNOR dots in interphase nuclei, can change even though the *actual number of NORs* in the nucleus is constant, AgNOR counts are not equivalent to NOR counts. This is because of changes in nucleolar structure during the cell cycle. Thus an increase in the mean number of AgNOR dots can be the result of an increased number of proliferating cells within a lesion.

The AgNOR number and area can also change as a result of fluctuations in transcriptional activity. With increased transcriptional activity, there is increased synthesis of argyrophilic NORAPs.

Proliferation or Ploidy?

Ploidy may influence AgNOR counts, but, in theory, only if the additional chromosomal material bears NOR sites. In aneuploid tumour nuclei, the higher AgNOR counts may, however, be due to concomitant proliferative activity, transcriptional activity or defective nucleolar association.

An elegant study to determine whether ploidy or proliferation influences AgNOR counts was performed on trophoblastic tissue by Suresh et al. (1990). Trophoblastic lesions are ideal for such a study because they exhibit a range of known ploidy and proliferative characteristics: hydropic abortions are diploid and lack proliferative activity; complete hydatidiform moles are diploid but proliferative; and partial hydatidiform moles are triploid, with less proliferation. AgNOR counts in partial moles were found to be approximately 50% higher than in the diploid categories of trophoblastic lesion. However, it is acknowledged that in neoplasms the effect of ploidy on AgNOR counts may be obscured by the greater influence of higher levels of proliferation.

Cytogenetic analysis of the relationship between AgNOR counts and ploidy has shown that, in non-Hodgkin's lymphomas, the number of NOR-bearing chromosomes does not necessarily correlate with the number of AgNOR dots in interphase nuclei (Jan-Mohamed et al. 1989). This discrepancy highlights the relationship between NORs and AgNORs, and suggests that other factors are responsible for the higher AgNOR counts witnessed in malignant lesions.

The principal factor responsible for increased AgNOR counts in neoplastic cells is the rate of cellular proliferation (Field et al. 1984). In tumour cell cultures, for example, there is correlation between AgNOR counts and AgNOR area (Derenzini et al. 1989, 1990) and the rate of proliferation. This is further substantiated by studies on neoplastic lesions showing a strong positive correlation between AgNOR scores and proliferation indices such as Ki-67 immunostaining.

Correlation with Other Indices of Proliferation

Comparison with DNA cytometric data and immunostaining indices for proliferation antigens (e.g. Ki-67) suggests that AgNOR counts in neoplastic cells are at least partly a measure of proliferative activity. Lesions with a high proportion of cells in S phase or high Ki-67 staining indices tend to have high AgNOR scores. These observations have promoted the AgNOR technique as a relatively simple, if time-consuming, alternative to the more complex methods. Unlike DNA cytometry, no expensive or sophisticated equipment is required other than that to be found in a routine histopathology department. Unlike Ki-67 immunostaining, the method can be applied to sections of routinely fixed and embedded tissue. Immunostaining for proliferation antigens, such as PCNA, surviving routine tissue processing may supersede the AgNOR technique simply as a measure of proliferation, because it does not involve tedious dot counting. However, recent work suggests that deregulated expression of PCNA in some malignant neoplasms may invalidate its use as a proliferation marker in some lesions (Chap. 8; Hall et al. 1990).

A few studies have sought correlation between AgNOR scores and other proliferation indices in human neoplasms.

Non-Hodgkin's Lymphomas

Crocker and Nar (1987) reported that mean AgNOR counts facilitate the distinction between high- and low-grade non-Hodgkin's lymphomas; the counts ranged from 4.4 to 6.8 and from 1.0 to 1.5 in the two groups, respectively. DNA flow cytometry was performed to determine, by correlation, the relative contributions of DNA ploidy and proliferation to the differences in AgNOR counts between the two categories of non-Hodgkin's lymphoma (Crocker et al. 1988). No correlation was found between mean AgNOR counts and DNA ploidy. However, the correlation between mean AgNOR counts and the S-phase fraction was strong ($r = 0.86$).

A similar correlation between mean AgNOR counts and proliferative status, determined by the Ki-67 staining index, was found by Hall et al. (1988). In a study of 35 low-grade and 45 high-grade non-Hodgkin's lymphomas, the relationship was highly significant ($r = 0.858$, $P < 0.001$; Fig. 9.5).

Simultaneous demonstration of AgNORs and Ki-67 staining in frozen sections enables the correlation to be established at a cellular level. In normal lymphoid tissue, Ki-67 positive nuclei tend to have higher AgNOR scores (Murray et al. 1989).

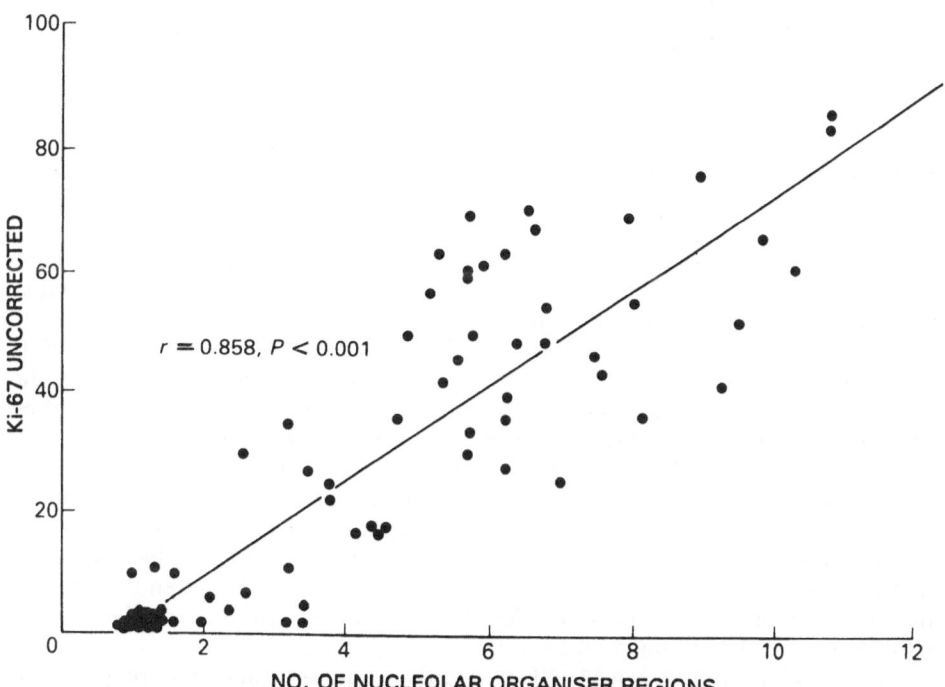

Figure 9.5. Ki-67 score, as an index of proliferation, correlated with mean number of nucleolar organiser regions per nucleus in 80 cases of non-Hodgkin's lymphoma. (From Hall et al. (1988), with permission.)

Breast Carcinoma

Dervan et al. (1989) assessed the AgNOR technique for estimating the proliferative state of benign and malignant breast lesions. The proliferative state was deduced from Ki-67 immunostaining indices. In benign lesions, the mean AgNOR count per cell was 2.65–6.80, contrasting with a range of 4.6–26.9 in carcinomas. There was a highly significant correlation ($P < 0.001$) between AgNOR counts and Ki-67 scores. It was concluded that AgNOR counts correspond to the proliferative state of the cells in the lesions studied. A similar conclusion was reached by Raymond and Leong (1989) and by Canepa et al. (1990).

Using DNA flow cytometry, Giri et al. (1989) reported an insignificant correlation between growth-phase fraction and AgNOR count. Breast carcinomas with a mean AgNOR count of more than 3 per nucleus were, however, more likely to be aneuploid than those with a mean AgNOR count of less than 3.

Lung Carcinomas

In contrast to observations on non-Hodgkin's lymphomas and some studies on breast carcinoma, no correlation between AgNOR scores and Ki-67 staining indices was found in a study of 95 lung tumours of various histological types (Soomro and Whimster 1990).

Applications in Histopathology

Ploton et al. (1986) showed that prostatic carcinoma nuclei tended to have larger nucleoli, harbouring numerous AgNOR dots, contrasting with smaller nucleoli and fewer dots in benign tissue. Crocker and his colleagues, and others, have extended these observations and have advocated use of the AgNOR technique in diagnostic histopathology for two main purposes – to assist in the distinction between benign and malignant lesions, and to assist in the grading of malignant neoplasms. Details can be found in Crocker's recent comprehensive review (Crocker 1990) and will not be reiterated here.

In general, the AgNOR technique appears to have some utility in tumour grading; high-grade tumours tend to have higher AgNOR counts than those lesions that pursue a less aggressive or benign clinical course. This applies to non-Hodgkin's lymphomas and to most carcinoma types other than thyroid and prostate; the association between AgNOR count and grade is relatively weak with lesions of the stomach and cervix.

The correlation between AgNOR counts, tumour grade and prognosis is almost certainly attributable to proliferation. Ploidy variations probably have a lesser influence. The effect of any defect of nucleolar association in neoplastic cells awaits elucidation.

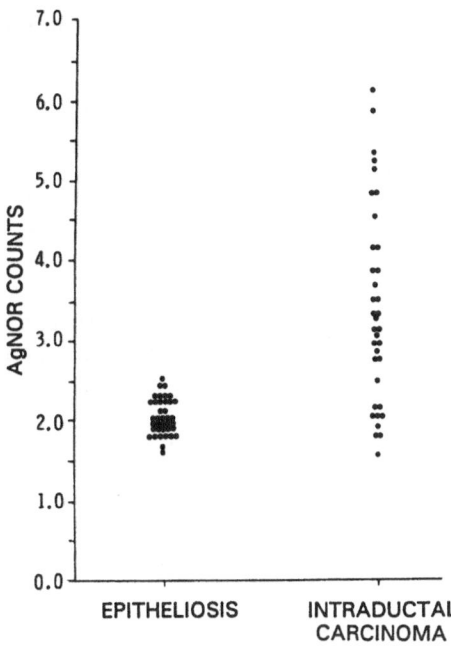

Figure 9.6. Mean AgNOR count per nucleus in cases of epitheliosis (ductal epithelial hyperplasia) and intraduct carcinoma of the breast. There is overlap in the region of 1.5–2.5 AgNORs per nucleus. (From Giri et al. (1989), with permission.)

The attraction of the AgNOR technique is its *apparent* objectivity; lesions are graded or classified on the basis of a numerical count rather than the subjective interpretation of histological features. However, an element of subjectivity is involved in interpreting clustered AgNORs.

As a sole criterion of malignancy in borderline cases, AgNOR counts are unlikely to win much favour because of the problem of overlap between benign and malignant categories (Fig. 9.6; Underwood and Giri 1988). Cases that are borderline on routine light microscopy often fall into the overlap region of AgNOR counts. Having made the diagnosis of malignancy, however, AgNOR counts are useful in tumour grading. Although the relationship between AgNOR counts and cell proliferation is less direct than, for example, the BUdR incorporation or ^{3}H-thymidine uptake, the simplicity of the technique is appealing and likely to continue to provide fruitful opportunities for research in tumour pathology.

Conclusion

Proteins associated with transcriptionally active nucleolar organiser regions can be demonstrated in interphase nuclei by a relatively simple silver-staining technique. The resulting tiny black dots – AgNORs – can be counted and the mean nuclear AgNOR count estimated. The mean AgNOR count in malignant neoplastic cell nuclei tends to be higher than in the corresponding benign or reactive lesions. AgNOR counts have, therefore, potential value in tumour histopathology.

Recent studies show that there is a correlation between AgNOR counts and cell proliferation indices such as mitotic counts and Ki-67 scores. However, AgNOR counts may not be a perfectly reliable paradigm for other measures of proliferative activity because the counts are influenced by other factors such as variations in ploidy, transcriptional activity, and possibly defective nucleolar association. Nevertheless, the clinical utility of AgNOR counts in tumour grading is worthy of further study.

Acknowledgement. The author is grateful to Dr Dilip Giri, a Commonwealth Medical Research Scholar, and to Dr Alec Howat for the benefit of their collaboration in studies of nucleolar organiser regions.

References

Alberts B, Bray D, Lewis J, Raff M, Roberts K and Watson JD (1989) Molecular biology of the cell, 2nd edn. Garland, New York, pp 539–546

Boldy DAR, Crocker J and Ayres JG (1989) Application of the AgNOR method to cell imprints of lymphoid tissue. J Pathol 157:75–79

Busch H, Daskal Y, Gyorkey F and Smetana K (1979) Silver staining of nucleolar granules in tumour cells. Cancer Res 39:857–863

Canepa M, Gambini C, Sementa AR, Borgiani L and Rovida S (1990) Nucleolar organizer regions and Ki-67 immunostaining in ductal breast cancer: a comparative study. Patologica 82:125–132

Chiu KY, Loke SL and Wong KK (1989) Improved silver technique for showing nucleolar organiser regions in paraffin wax sections. J Clin Pathol 42:992–994

Coghill G, Grant A, Orrell JM, Jankowski J and Evans AT (1990) Improved silver staining of nucleolar organiser regions in paraffin wax sections using an inverted incubation technique. J Clin Pathol 43:1029–1031

Crocker J (1990) Nucleolar organiser regions. In: Underwood JCE (ed) Pathology of the nucleus. Curr Top Pathol 82:91–149

Crocker J and Nar P (1987) Nucleolar organiser regions in lymphomas. J Pathol 151:111–118

Crocker J and Egan MJ (1988) Correlation between NOR sizes and numbers in non-Hodgkin's lymphomas. J Pathol 156:233–239

Crocker J, Macartney JC and Smith PJ (1988) Correlation between DNA flow cytometric and nucleolar organiser region data in non-Hodgkin's lymphomas. J Pathol 154:151–156

Crocker J, Boldy DAR and Egan MJ (1989) How should we count AgNORs? Proposals for a standardized approach. J Pathol 158:185–188

Derenzini M, Pession A, Farabegoli F, Badiali M and Dehan P (1989) Relationship between interphasic nucleolar organiser regions and growth rate in two neuroblastoma cell lines. Am J Pathol 134:925–932

Derenzini M, Pession A and Trere D (1990) Quantity of nucleolar silver-staining proteins is related to proliferating activity in cancer cells. Lab Invest 63:137–140

Dervan PA, Gilmartin LG, Loftus BM and Carney DN (1989) Breast carcinoma kinetics: argyrophilic nucleolar organizer region counts correlate with Ki-67 scores. Am J Clin Pathol 92:401–407

Field DH, Fitzgerald PH and Sin FYT (1984) Nucleolar silver-staining patterns related to cell cycle phase and cell generation of PHA-stimulated human lymphocytes. Cytobios 41:23–33

Giri DD, Nottingham JF, Lawry J, Dundas SAC and Underwood JCE (1989) Silver-binding nucleolar organizer regions (AgNORs) in benign and malignant breast lesions: correlations with ploidy and growth phase by DNA flow cytometry. J Pathol 157:307–313

Goodpasture C and Bloom SE (1975) Visualisation of nucleolar organiser regions in mammalian chromosomes using silver staining. Chromosoma 53:37–50

Griffiths AP, Butler CW, Roberts P, Dixon MF and Quirke P (1989) Silver-stained structures (AgNORs): their dependence on tissue fixation and absence of prognostic relevance in rectal adenocarcinoma. J Pathol 159:121–127

Hall PA, Crocker J, Watts A and Stansfeld AG (1988) A comparison of nucleolar organiser region staining and Ki-67 immunostaining in non-Hodgkin's lymphoma. Histopathology 12:373–381

Hall PA, Levison DA, Woods AL et al. (1990) Proliferating cell nuclear antigen (PCNA) immunolocalisation in paraffin sections: an index of cell proliferation with evidence of deregulated expression in some neoplasms. J Pathol 162:285–294

Howell WM (1977) Visualisation of ribosomal gene activity: silver stains proteins associated with rRNA transcribed from oocyte chromosomes. Chromosoma 62:361–367

Jan-Mohamed RM, Armstrong SJ, Crocker J, Leyland MJ and Hulten MA (1989) The relationship between number of interphase NORs and NOR-bearing chromosomes in non-Hodgkin's lymphoma. J Pathol 158:3–7

Murray PG, Boldy DAR, Crocker J and Ayres JG (1989) Sequential demonstration of antigens and AgNORs in frozen and paraffin sections. J Pathol 159:169–172

Ploton D, Menager M, Jeannesson P, Himber G, Pigeon J and Adnet J-J (1986) Improvement in the staining and in the visualisation of the argyrophilic proteins of the nucleolar organizer region proteins at the optical level. Histochem J 18:5–14

Quinn CM and Wright NA (1990) The clinical assessment of proliferation and growth in human tumours: evaluation of methods and applications as prognostic variables. J Pathol 160:93–102

Raymond WA and Leong AS-Y (1989) Nucleolar organiser regions relate to growth fractions in human breast carcinoma. Hum Pathol 20:741–746

Rüschoff J, Plate KH, Contractor H, Kern S, Zimmermann R and Thomas C (1990) Evaluation of nucleolus organizer regions (NORs) by automatic image analysis: a contribution to standardization. J Pathol 161:113–118

Smith PJ, Skilbeck NQ, Harrison A and Crocker J (1988) The effect of a series of fixatives in the AgNOR technique. J Pathol 155:109–112

Soomro IN and Whimster WF (1990) Growth fraction in lung tumours determined by Ki-67 immunostaining and comparison with AgNOR scores. J Pathol 162:217–222

Suresh UR, Chawner L, Buckley CH and Fox H (1990) Do AgNOR counts reflect cellular ploidy or cellular proliferation? A study of trophoblastic tissue. J Pathol 160:213–215

Underwood JCE and Giri DD (1988) Nucleolar organizer regions as diagnostic discriminants for malignancy. J Pathol 155:95–96

10 Clinical Aspects of Assessing Cell Proliferation

S.M. O'REILLY and M.A. RICHARDS

Introduction

Despite the numerous studies which have assessed cell proliferation in tumours, measurement of proliferation indices has had, to date, little impact on patient management. This chapter will outline the techniques currently clinically available and discuss their practical advantages and disadvantages. In addition, the clinical situations in which assessment of cell proliferation may prove to be of use will be reviewed.

Methods Used to Assess Cell Proliferation

The individual techniques currently used to assess cell proliferation are discussed in detail in other chapters. The aim of this section is to examine the relative advantages and disadvantages of these methods in clinical practice, a topic which has been reviewed recently (Hall and Levison 1990; Quinn and Wright 1990). Firstly, it is important to note that the various methods used to estimate cell proliferation actually attempt to measure different parameters. Thus, the results of studies addressing the same question but using different approaches to measure proliferation may not be directly comparable (Silvestrini et al. 1988a). Even techniques such as FCM and TLI, which purport to estimate the same phenomenon – the SPF – can give markedly different results (Dressler and Bartow 1989), while in the case of AgNORs and Ki-67 we remain uncertain of the exact nature of the biological process being measured.

Techniques that rely on manual counting – whether of mitoses, AgNORs, cells labelled with thymidine or BUdR or immunostained to detect proliferation-associated antigens – are all subject to a number of inherent methodological problems. They are time-consuming procedures which are prone to inter-observer variation and sampling errors. There is a lack of consensus on how many cells need to be counted to produce a valid result (Hall and Levison 1990; Quinn and Wright 1990). Some techniques – thymidine labelling, BUdR labelling and Ki-67 staining – can only be performed on fresh or frozen tissue. Such methods do, however, have the advantage of retaining the spatial orientation of the tissue and of enabling the tumour cells to be distinguished from stromal cells.

DNA FCM is an objective method which allows rapid analysis of large numbers of cells and can be performed on both fresh and histologically processed tissue. However, it requires physical disruption of tissues, and the results can be influenced by the dilutional effect of non-tumour cells. In addition, there is no consensus on how best to estimate proliferative capacity from DNA histograms. Numerous mathematical models have been developed, but all have difficulty in estimating the SPF in tumours with overlapping diploid and aneuploid cell populations (Baisch et al. 1982; Dean and Jett 1974; Dressler et al. 1987).

Finally, there are two further difficulties concerning clinical assessment of cell proliferation, irrespective of the method used. Firstly, all rely on the analysis of only a small sample of the tumour cells, even though there is good evidence that tumours can display a marked heterogeneity of growth patterns. This therefore limits the validity of using a single sample as an index of proliferative capacity (Aherne et al. 1977; McFarlane et al. 1986). Secondly, the methods in common use usually give only a "snapshot" assessment of the percentage of cells actively in cycle at one moment. No account is taken of the rate at which cells enter the cycle, the duration of its various phases, or the rate of cell loss from the proliferative compartment through differentiation or cell death.

Clinical Applications

Given all the difficulties in interpreting the results of kinetic studies discussed above, it is not surprising that there is still much debate on how meaningful such measurements are in the clinical setting. In order to examine this question, the evidence that measurements of cell proliferation can give information on diagnosis, prognosis, selection of patients for treatment and their response to therapy will be reviewed.

Diagnosis

While histopathologists have long taken mitotic activity into account when making the diagnosis of malignancy, formal measurement of proliferative activity is not generally used to distinguish benign from malignant lesions. One possible exception to this is the differential diagnosis of smooth muscle tumours of the uterus. While the pathological distinction between a clinically benign tumour and its malignant counterpart is usually straightforward, a minority of cases prove difficult because of hypercellularity and nuclear atypia. Taylor and Norris (1966) advocated the use of mitosis counting in distinguishing leiomyoma from leiomyosarcoma. They noted that 27 tumours with fewer than 10 mitoses per 10 HPFs, which had previously been classified as leiomyosarcoma, followed a benign course, while 31 of 36 tumours with 10 or more mitoses per 10 HPFs metastasised or recurred locally. Several other authors have confirmed the importance of mitotic count in diagnosing leiomyosarcoma, although they differ on the number of mitoses needed per HPF to make a diagnosis of malignancy (Christopherson et al. 1972; Ellis and Whitehead 1981). The use of the mitotic count has been criticised, however, because of 600% variation between different microscopes in the area of the HPF, and MI (the fraction of cells containing mitotic figures) has been recommended as a preferable method (see Chap. 5; Ellis and Whitehead 1981).

There is considerable overlap between the proliferative activity of tumour cells and that of regenerating but non-neoplastic tissue. Indeed there may be considerably more proliferation in reactive processes than in neoplasms, as exemplified by reactive germinal centres and follicular lymphomas. Such kinetic heterogeneity limits the usefulness of measurements of proliferation in distinguishing benign from malignant lesions (Schrape et al. 1987). There are problems also in using DNA ploidy to distinguish malignant from benign lesions. Although aneuploidy has been reported in approximately 75% of solid tumours (Barlogie et al. 1983), the presence of aneuploidy does not necessarily imply malignancy. For example, aneuploidy has been reported in 30% of benign prostate specimens (De Vere White et al. 1987) and premalignant lesions such as colorectal adenomas can have aneuploid populations of cells (Sciallero et al. 1988). In problematic areas such as thyroid follicular neoplasms, where the distinction between follicular adenomas and follicular carcinomas can be difficult histologically, the presence of aneuploidy in up to 25% of adenomas means that the determination of DNA ploidy gives little diagnostic assistance to the pathologist (Grant et al. 1990). The presence of aneuploidy may have a role, however, in the detection of bladder tumours. Collste et al. (1980), using the presence of either aneuploid cells or a high (>15%) percentage of cells in the $S+G_2/M$ phase in bladder washings to diagnose carcinoma, successfully identified 34/38 histologically proven tumours. FCM of bladder washings has been shown to be more effective at detecting superficial bladder cancer than either voided urine or irrigation cytology (Badalament et al. 1987).

The demonstration of NORs (see Chap. 9) has been claimed to be of diagnostic value in breast (Smith and Crocker 1988), lung (Crocker et al. 1987), prostate (Ploton et al. 1986) cutaneous (Crocker and Skilbeck 1987; Fallowfield et al. 1988) and gastrointestinal tract malignancies (Yang et al. 1990), and in some small round cell tumours of childhood (Egan et al. 1987). In most studies, however, the significant differences reported have been between clear-cut benign and malignant lesions and the wide scatter of AgNOR counts observed has led to questioning of the value of such measurements in the differential diagnosis of borderline malignant lesions (Walker 1988). Recent studies have indicated that AgNOR counts are unlikely to be helpful in distinguishing regenerative epithelium from early gastric malignancy (Suarez et al. 1989), florid epitheliosis from intraduct breast carcinoma (Giri et al. 1989), small cell carcinoma of the bronchus from bronchial carcinoid (Benbow and Cromie 1989), or borderline melanocytic lesions from melanoma (Fallowfield and Cook 1989).

If indices of cell proliferation are seldom of help in the diagnosis of malignancy, can they be used to identify patients at high risk of developing a
tumour? It has been proposed that measurement of thymidine incorporation
into colonic mucosa may be of use in the follow-up of patients with chronic
inflammatory bowel disease and colorectal adenoma, conditions associated
with an increased incidence of large bowel cancer. Lipkin, Deschner and co-
workers (1983) used in vitro tritiated thymidine labelling to study endoscopic
biopsies from the normal-appearing, flat rectal mucosa of control subjects
and of patients with familial polyposis coli. They noted that, while the rectal
mucosa of normal subjects is characterised by a high kinetic activity only in
the lower portion of the crypts, in patients with familial polyposis coli proliferating cells are present throughout the whole length of the crypt. A similar
abnormal pattern of cell proliferation has also been found along the entire
colon in patients with colon cancers (Terpstra et al. 1987). Similar studies on
patients with extensive ulcerative colitis in remission showed no increase in
overall labelling index but, as noted in patients with adenomas and carcinomas, did show a significant upward displacement of the proliferative zone
compared to normal controls (Biasco et al. 1990). This was not related to
the duration of colitis, and it showed a bimodal distribution, all patients with
frank epithelial dysplasia being in the group with the higher values. While
the authors postulate that such measurements could help select a subgroup
of patients on whom to intensify surveillance, it is difficult to see how such
a relatively sophisticated and time-consuming assessment of proliferation
could be widely applied. The application of immunohistological methods
(Chap. 8) might, however, be practical.

Prognosis

It is as a prognostic variable that clinical assessment of cell proliferation in
tumours has been most extensively investigated. With more than 100 articles
on the prognostic value of DNA FCM in 1989 alone, it is not possible to review all the published data in one chapter. We will therefore use breast cancer
and non-Hodgkin's lymphoma, two of the tumour types which have been most
exhaustively studied, to illustrate some general principles. The possible role
of kinetic indices as prognostic factors in other tumour types will be briefly
summarised.

Lymphoma

Examination of the relationship between indices of cell proliferation and prognosis in non-Hodgkin's lymphoma (NHL) is complicated by a number of confounding variables. Even when the possible differences between the various
methods used to assess proliferation are disregarded, there are pathological
and clinical factors which make comparison between studies difficult. The
problems associated with the histological classification of NHL are well recognised (NCI non-Hodgkin's lymphoma Classification Project 1985). Even when
pathologists agree on which classification should be used, there is evidence
of considerable inter-observer difference in the grading of tumours (Hanby

et al. 1990). Series vary also as to whether other prognostic variables such as stage, tumour bulk and the presence of B symptoms are taken into account when assessing the prognostic value of kinetic indices. Finally, a great variety of treatments have been used, which may well have different impacts on outcome and thus partially abrogate the effect of other prognostic variables such as kinetic indices.

Bearing these problems in mind, what information is currently available about the prognostic value of assessment of cell proliferation in NHL? Thymidine labelling (Brandt et al. 1981; Costa et al. 1981), FCM (Diamond et al. 1982; Rehn et al. 1990) Ki-67 immunostaining (Hall et al. 1988a; Schrape et al. 1987) and AgNOR staining (Crocker and Nar 1987; Hall et al. 1988a) have, in general, demonstrated a strong association between a high level of proliferative activity and a high tumour grade, but there is substantial kinetic heterogeneity within and between histological subtypes. Given the close relationship between grade and clinical outcome for patients with NHL (Richards et al. 1987), it is not surprising that in several studies a relationship between high proliferative activity and poor outcome has been demonstrated. Of greater interest is the possibility that measurements of proliferative activity may be of prognostic significance within tumour grades and may thus provide information additional to that provided by routine histopathology. There is evidence, using thymidine labelling (Kvalov et al. 1985), FCM (Egerton et al. 1988; Rehn et al. 1990) and Ki-67 (Hall et al. 1988a) to assess proliferation, that patients with low-grade NHL whose tumours have a high proliferative index have a worse prognosis than patients whose tumours are of similar histology but with a low proliferative index. This effect of proliferative activity was, in addition to the presence of B symptoms, one of the strongest predictors of survival in one multivariate analysis (Rehn et al. 1990). There is not universal agreement, however, on the usefulness of such measurements in low-grade NHL; another study, on a small number of patients with follicular lymphomas, suggested that Ki-67 immunostaining and AgNOR counting offer no advantage over morphological assessment in identifying patients with a poor short-term prognosis (Cibull et al. 1989). In addition to the possible effect of proliferative indices on survival, it has been suggested that estimation of SPF by FCM may help identify patients with low-grade NHL who are at risk of transformation to high grade disease (Macartney et al. 1986).

In high-grade lymphoma, the relationship between proliferative activity and clinical outcome is less clear. The impact of the results of DNA FCM on prognosis in NHL has recently been reviewed (Macartney and Camplejohn 1990). Ploidy status by itself was felt to have little effect on overall survival. Several workers have noted an association between low proliferative activity and longer survival (Bauer et al. 1986; Srigley et al. 1985; Wooldrige et al. 1988), but others have found no such relationship (Cowan et al. 1989; Rehn et al. 1990). Nonetheless, Macartney and Camplejohn (1990) concluded that there was a strong suggestion that high proliferative activity adversely affects survival, at least in the short term, but emphasised the urgent need for further studies on well defined groups of cases. A similar association between high proliferative activity and shorter survival has also been suggested from use of thymidine labelling to assess tumour kinetics (Costa et al. 1981). In high-grade NHL, Ki-67 immunostaining has also been assessed as a prognostic factor. Hall et al. (1988b) reported that a high Ki-67 index was not an adverse prognostic factor in high-grade lymphoma. Indeed, they observed that, for patients achieving remission with first-line chemotherapy, a very high Ki-67 index (>80%) was associated with a lower risk of relapse. This observation might be explained by a small proportion of tumour cells escaping the effects of drug treatment by being in the resting phase of the cycle (G_0) during each cycle of treatment. Thus, despite an apparent good response to therapy, resistant clones of tumour cells regrow leading to relapse.

Breast Cancer

A variety of clinicopathological variables, including the presence of axillary lymph node metastases (Fisher et al. 1983), tumour size (Nemoto et al. 1980), tumour grade (Bloom and Richardson 1957) and steroid receptor status (Mason 1983), have been shown to provide important prognostic information in patients with breast cancer. Many centres have examined the correlations between these tumour features and kinetic indices. In studies using thymidine labelling, proliferative activity has not been significantly correlated with lymph-node status (Gentili et al. 1981; Meyer and Hixon 1979; Meyer et al. 1986; Tubiana and Malaise 1976) and, while a significant trend for TLI to increase with increasing tumour diameter has been reported (Meyer et al. 1986), most studies have noted only a weak (Meyer and Hixon 1979; Meyer et al. 1986) or no (Courdi et al. 1989; Tubiana and Malaise 1976) association between TLI and size. In contrast, a strong association between high TLI and high tumour grade has been consistently demonstrated (Courdi et al. 1989; Gentili et al. 1981; Holt et al. 1986; Tubiana et al. 1981). Generally, an inverse relationship has also been observed between TLI and both oestrogen receptor status (Bertuzzi et al. 1981; Courdi et al. 1989; Meyer et al. 1986; Straus et al. 1982) and progesterone receptor status (Courdi et al. 1989; Meyer et al. 1986), although this association has often been weak.

The results of studies examining the relationship between proliferative activity estimated by FCM and other features of breast tumours are summarised in Table 1. Once again, there is a consistent and strong association between high proliferative activity and poor tumour grade, a weaker correlation with steroid receptor status, and little evidence of a relationship with tumour size or nodal status. A similar pattern of associations has also been reported using Ki-67 (Barnard et al. 1987; Gerdes et al. 1987; McGurrin et al. 1987) and AgNORs (Giri et al. 1989) to assess proliferation.

The majority of studies have shown an association between high proliferative activity and a poor prognosis using either TLI or FCM to assess proliferation, with few exceptions (Holt et al. 1986; Lykkesfeldt et al. 1988). Of greater interest is whether kinetic indices retain their prognostic significance when other features of the tumour are included in a multivariate analysis. In general, while proliferative activity remains a significant independent predictor of outcome when size, nodes, ploidy and steroid receptor status are entered into the analysis (Clark et al. 1989; Klintenberg et al. 1986; Sigurdsson et al. 1990; Stal et al. 1989; Tubiana et al. 1981), this significance was lost when tumour grade is included (Hedley et al. 1987; O'Reilly et al. 1990a). However, Toik-kanen et al. (1989) have reported tumour grade and SPF to be independent predictors of survival.

Table 10.1. Relationship between proliferative activity (measured by DNA FCM) and other features of breast cancer

Reference	No.	Nodes	Size	Grade	Oestrogen receptor (ER) status	Progesterone receptor (PgR) status
Olszewski et al. (1981)	90	NA	0	+	+	NA
Raber et al. (1982)	80	0	NA	NA	+	NA
Kute et al. (1981)	70	0	NA	NA	+	+*
Moran et al. (1984)	104	0	NA	++	++	+*
Haag et al. (1984)	155	NA	0	NA	NA	NA
Fossa et al. (1984)	66	NA	NA	++	NA	NA
McDivitt et al. (1985)	75	NA	NA	NA	++	0
Kute et al. (1985)	179	0	0	++	++	+
McDivitt et al. (1986)	168	0	0	++	++	+
Kallioniemi et al. (1987)	59	NA	NA	++	+	++
Hedley et al. (1987)	285	all N+	0	++	0	NA
Dressler et al. (1987)	1084	+†	NA	NA	++	++*
Feichter et al. (1988)	300	0	0	++	++	0
Stal et al. (1989)	290	0	++	NA	++	NA
Muss et al. (1989)	84	all N0	0	++	+	+
Toikkanen et al. (1989)	223	+‡	0	++	NA	NA
Christov et al. (1989)	180	NA	NA	++	NA	NA
Sigurdsson et al. (1990)	367	all N0	+	NA	+	+
O'Reilly et al. (1990a)	134	0	0	++	0	0

Abbreviations: 0, no significant association; +, weak association $(0.01 < P < 0.05)$; ++, strong association $(P < 0.01)$; NA, not assessed.
* ER+ve PgR+ve versus ER−ve PgR−ve.
† Significant association within diploid tumours only.
‡ Nodes assessed clinically.

With the recent suggestion that adjuvant chemotherapy may have a role in the management of patients with node-negative breast cancer, there has been particular interest in the ability of kinetic indices to help define prognostic groups within node-negative disease. The Milan group have analysed data from more than 400 patients and have observed a significantly shorter relapse-free survival and survival for patients whose tumours have high TLI (Gentili et al. 1981; Silvestrini et al. 1985; Silvestrini et al. 1988b). Similar results have also been noted by other groups (Courdi et al. 1989; Meyer et al. 1983). An association between high SPF and poor prognosis in node-negative breast cancer has also been observed in studies using FCM to estimate proliferation (Clark et al. 1989; Dressler et al. 1990; O'Reilly et al. 1990b; Sigurdsson et al. 1990), although Clark et al. (1989) found SPF to have prognostic significance only for patients with diploid tumours. In several reports the predictive value of SPF was still evident after multivariate analysis (Clark et al. 1989; O'Reilly et al. 1990b; Sigurdsson et al. 1990). O'Reilly et al. (1990c) combined tumour size and SPF to identify three prognostic subgroups with 5-year relapse-free survivals of 96%, 78% and 52%, respectively (Figure 10.1). Such prognostic groupings may help select patients for adjuvant therapy.

Other Tumours

The great majority of studies examining the prognostic importance of proliferative activity in other tumour types have involved the use of DNA FCM

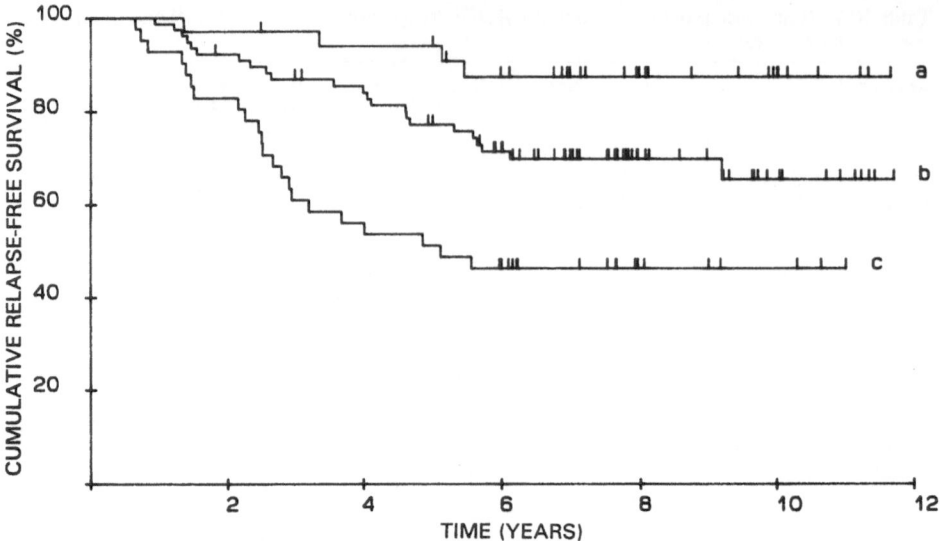

Figure 10.1. Relapse-free survival for patients with node-negative breast cancer: a, tumour ≤ 1 cm (n = 33), b, tumour > 1 cm plus low SPF (n = 76), c, tumour > 1 cm plus high SPF (n = 41).

or thymidine labelling. The data on the usefulness of DNA ploidy in this set-ting have recently been reviewed (Dressler and Bartow 1989; Merkel and McGuire 1990) and will not be discussed further. Tumour proliferative activity has been studied less exhaustively and, as with breast cancer and lymphoma, it is important to interpret the data in the light of other potential prognostic factors. In ovarian cancer, a high SPF has been reported to be associated with shorter survival, and this effect was still significant after adjusting for stage, residual tumour after surgery, histology, age and treatment used (Kallioniemi et al. 1988). The unfavourable impact of high SPF on survival was most notice-able for patients with diploid tumours and low (I/II) stage disease. An in-creased SPF has also been noted to be associated with a worsened prognosis in non-small cell lung cancer, independent of histology or stage of disease (Ten Velde et al. 1988; Volm et al. 1985), although this observation has not been universal (Bunn et al. 1983).

There is also conflict over the prognostic role of proliferative activity in colorectal cancer. Survival was not influenced by SPF measured by thymidine labelling in one study (Meyer and Prioleau 1981). In contrast, Bauer et al. (1987) found high SPF, measured by FCM, to be a significant predictor of poor survival, although this association retained only marginal significance after correction for clinical stage and patient age. However, when stage-D disease was excluded from the analysis, SPF emerged as the only independent prog-nostic factor both in stage-A and -B disease, grouped together, and in stage-C disease. A similar effect of SPF, but confined to patients with stage-C disease, has also been observed by another group (Schutte et al. 1987).

The TLI of plasma cells has been shown to be useful in predicting prognosis in multiple myeloma. Even with aggressive treatment myeloma is not curable but many patients, both before and after therapy, can have stable or indolent disease which does well without any specific treatment (Alexanian et al. 1978;

Durie et al. 1980). Conditions that can remain clinically stable for many years – such as monoclonal gammopathy of undetermined significance, smouldering or indolent myeloma – have extremely low labelling indices of plasma cells, while patients with active myeloma have significantly higher plasma cell labelling indices (Boccadoro et al. 1984; Durie et al. 1980; Greipp and Kyle 1983). In addition, a low labelling index was also noted in patients in stable remission following induction chemotherapy (Boccadoro et al. 1984). Thus, low TLI can help identify patients who can be safely followed without treatment (Durie 1986).

Predicting Response to Treatment

Compared with the multitude of studies examining the ability of proliferative activity to predict relapse-free survival and survival, there are relatively few reports which analyse directly the relationship between kinetic indices and response to treatment. Data from studies addressing this question in high- and intermediate-grade NHL are conflicting. One study noted an increased response rate in tumours with higher proliferative indices (Srigley et al. 1985), others have reported a significant association between low proliferative index and induction of response (Cowan et al. 1989; Wooldrige et al. 1988) and yet others have found no significant correlation between proliferative activity and response (Bauer et al. 1986; Winter et al. 1989). As with the prognostic implications of proliferative activity, these results are difficult to interpret because of the heterogeneity of tumour subtypes and of treatments used.

The association between proliferative activity and response to therapy has been examined for patients with breast cancer in a number of clinical settings although, once again, there have been relatively few studies. Remvikos et al. (1990) have demonstrated a relationship between high SPF, measured by FCM on a fine-needle aspirate specimen, and response of the unresected primary tumour to neoadjuvant chemotherapy. In a series of 89 patients with stage-II disease objective response rates increased from 41% for tumours with an SPF of less than 5% to 94% for tumours with more than 10% cells in SPF ($P < 0.0001$). The response of occult micrometastases to adjuvant chemotherapy has also been assessed, although in this situation response to treatment can only be measured indirectly as a prolongation of the time to relapse. Bonadonna et al. (1986) have suggested that the major impact of adjuvant chemotherapy in prolonging relapse-free survival in patients with node-negative breast cancer occurs in patients whose tumours have a high TLI, with little benefit seen in those whose tumours have a low TLI. In contrast, in a study conducted at Guy's hospital on 214 patients with node-positive breast cancer randomised to receive adjuvant CMF or to a control group, chemotherapy significantly improved relapse-free survival, both for premenopausal patients whose tumours had a low SPF (measured by FCM) and for premenopausal patients whose tumours had a high SPF (O'Reilly et al. 1990c). No significant benefit attributable to adjuvant chemotherapy was seen for either subgroup of postmenopausal patients.

The impact of the proliferative activity of primary breast tumours on the response of metastatic disease to therapy has also been analysed. One small study of 25 patients, using TLI, has shown an association between more rapid tumour proliferation and response to a variety of chemotherapy regimens (Sulkes et al. 1979). However, another study of 76 patients treated with adriamycin ± vincristine found no correlation between either DNA ploidy or SPF measured by FCM and response to chemotherapy (Masters et al. 1987). With regard to endocrine therapy for metastatic breast cancer, Stuart-Harris et al. (1985) found no significant difference in response between diploid and aneuploid groups in a study of 42 patients. More recently, a study of 136 patients treated with tamoxifen as first-line endocrine therapy reported a significantly higher response rate in patients with tetraploid or near-tetraploid tumours (Baildam et al. 1987). However, this result is made difficult to interpret by the unusually high incidence (36%) of tetraploid tumours, the inclusion of patients with stable disease in the group defined as responders, and the fact that tetraploid tumours also had a significantly higher rate of oestrogen-receptor positivity.

In patients with advanced squamous cell carcinoma of the head and neck region, ploidy has been analysed as a prognostic indicator in patients randomised in two clinical trials to receive either cisplatinum or to act as untreated controls (Cooke et al. 1990). While there was a significant prolongation of survival noted for patients in both ploidy subgroups compared with their untreated controls, the increase in median survival of only 6 weeks for patients with diploid tumours compared with 6 months for patients with aneuploid tumours led the authors to cast doubt on the usefulness of chemotherapy for patients with diploid tumours. This apparent unresponsiveness of diploid tumours to cisplatinum chemotherapy has been confirmed in another report in which all the pre-treatment diploid tumours remained histologically malignant following therapy, while repeat biopsies from tumours that were initially aneuploid revealed only diploid, histologically normal tissue in 48% of cases (Ensley et al. 1990). A similar relationship between ploidy, SPF and response to radiotherapy for an earlier stage of disease has been reported in a small study by Franzen et al. (1986), who observed that aneuploid tumours responded better to radiotherapy and that the mean SPF was higher in those tumours eradicated by preoperative radiotherapy than in those that did not respond.

Conclusion

Despite the vast and ever-expanding literature on proliferative activity, there is almost no clinical situation in which such information is used on a routine basis. This is due, firstly, to the lack of simple, cheap, reproducible and easily applied methods to investigate tumour kinetics. All the techniques alluded to in this review fail to meet these criteria, and thus the investigation of kinetic indices has been confined to a number of specialised centres.

Leaving aside the difficulties associated with the measurement of cell proliferation, there is little clear-cut evidence, as yet, that such measurements should influence patient management. Apart from the counting of mitoses in smooth muscle tumours, kinetic indices are not used in the diagnosis of malignancy. While there is considerable evidence that assessing proliferative activity does provide information on prognosis, there is a great need for studies to examine the role of kinetic indices by multivariate analyses, taking other clinicopathological variables into account. Moreover, little attention has been paid, so far, to the relationship between tumour kinetics and response to treatment.

In summary, the number of techniques available for assessing proliferation has expanded greatly in the past few years. What is urgently needed is the application of these techniques in a reproducible way to well-defined groups of patients in order to provide information of clinical relevance.

References

Aherne WA, Al-Wiswazy M, Ford D and Jellerer AM (1977) Assessment of inherent fluctuations of mitotic and labelling indices of human tumours. Br J Cancer 36:577–591

Alexanian R, Gehan E, Haut A et al. (1978) Unmaintained remission in multiple myeloma. Blood 51:1005–1011

Badalament RA, Hermansen DK, Kimmel M et al. (1987) The sensitivity of bladder wash flow cytometry, bladder wash cytology and voided cytology in the detection of bladder carcinoma. Cancer 60:1423–1427

Baildam AD, Zaloudik J, Howell A et al. (1987) DNA analysis by flow cytometry, response to endocrine treatment and prognosis in advanced carcinoma of the breast. Br J Cancer 55:553–559

Baisch H, Beck H-P, Christensen IJ et al. (1982) A comparison of mathematical methods for the analysis of DNA histograms obtained by flow cytometry. Cell Tissue Kinet 15:235–249

Barlogie B, Raber MN, Schumann J et al. (1983) Flow cytometry in clinical cancer research. Cancer Res 43:3982–3997

Barnard NJ, Hall PA, Lemoine NR and Kadar N (1987) Proliferative index in breast carcinoma determined in situ by Ki-67 immunostaining and its relationship to clinical and pathological variables. J Pathol 152:287–295

Bauer KD, Merkel DE, Winter JN et al. (1986) Prognostic implications of ploidy and proliferative activity in diffuse large cell lymphomas. Cancer Res 46:3173–3178

Bauer KD, Lincoln ST, Vera-Roman JM et al. (1987) Prognostic implications of proliferative activity and DNA aneuploidy in colonic adenocarcinomas. Lab Invest 57:329–335

Benbow EW and Cromie CJ (1989) Inability of AgNOR counts to differentiate between bronchial carcinoid tumours and small cell carcinoma of the bronchus. J Clin Pathol 42:1003–1004

Bertuzzi A, Daidone MG, Di Fronzo G and Silvestrini R (1981) Relationship among estrogen receptors, proliferative activity and menopausal status in breast cancer. Breast Cancer Res Treat 1:253–262

Biasco G, Paganelli GM, Miglioli M et al. (1990) Rectal cell proliferation and colon cancer risk in ulcerative colitis. Cancer Res 50:1156–1159

Bloom HJG and Richardson WW (1957) Histological grading and prognosis in breast cancer. A study of 1409 cases of which 359 have been followed for 15 years. Br J Cancer 9:359–377

Boccadoro M, Gavarotti P, Fossati G et al. (1984) Low plasma cell ^{3}(H)thymidine incorporation in monoclonal gammopathy of undetermined significance (MGUS), smouldering myeloma and remission phase myeloma: a reliable indicator of patients not requiring therapy. Br J Haematol 58:689–696

Bonadonna G, Valagussa P and Tancini G et al. (1986) Current status of Milan adjuvant chemotherapy trials for node-positive and node-negative breast cancer. Natl Cancer Inst Monogr 1:45–49

Brandt L, Olsson H and Monti M (1981) Uptake of thymidine in lymphoma cells obtained through fine needle aspiration biopsy: relation to prognosis in non-Hodgkin's lymphoma. Eur J Cancer 17:1229–1233

Bunn PA, Carney DN and Gazdar AF et al. (1983) Diagnostic and biological implications of flow cytometric DNA content analysis in lung cancer. Cancer Res 43:5026, 5032

Christopherson WM, Williamson FO and Gray IA (1972) Leiomyosarcoma of the uterus. Cancer 29:1512–1517

Christov K, Milev A and Todorov V (1989) DNA aneuploidy and cell proliferation in breast tumours. Cancer 64:673–679

Cibull ML, Heryet A, Gatter KC and Mason DY (1989) The utility of Ki-67 immunostaining, nucleolar organizer region counting and morphology in the assessment of follicular lymphomas. J Pathol 158:189–193

Clark GM, Dressler LG and Owens MA et al. (1989) Prediction of relapse or survival in patients with node-negative breast cancer by DNA flow cytometry. N Engl J Med 320:627–633

Collste LG, Devonec M and Darzynkiewicz Z et al. (1980) Bladder cancer diagnosis by flow cytometry: correlation between cell samples from biopsy and bladder irrigation fluid. Cancer 45:2389–2394

Cooke LD, Cooke TG, Bootz F et al. (1990) Ploidy as a prognostic indicator in end stage squamous cell carcinoma of the head and neck region treated with cisplatinum. Br J Cancer 61:759–762

Costa A, Bonadonna G, Villa E et al. (1981) Labelling index as a prognostic marker in non-Hodgkin's lymphoma. J Natl Cancer Inst 66:1–5

Courdi A, Hery M, Dahan E et al. (1989) Factors affecting relapse in node-negative breast cancer: a multivariate analysis including labelling index. Eur J Cancer Clin Oncol 25:351–356

Cowan RA, Harris M, Jones M and Crowther D (1989) DNA content in high and intermediate grade non-Hodgkin's lymphoma – prognostic significance and clinicopathological correlations. Br J Cancer 60:904–910

Crocker J and Nar P (1987) Nucleolar organiser regions in lymphomas. J Pathol 151:111–118

Crocker J and Skilbeck N (1987) Nucleolar organiser region associated proteins in cutaneous melanotic lesions: a quantitative study. J Clin Pathol 40:885–889

Crocker J, Ayres J and McGovern J (1987) Nucleolar organiser regions in small cell carcinoma of the bronchus. Thorax 42:972–975

Dean PN and Jett JH (1974) Mathematical analysis of DNA distributions derived from flow microfluorometry. J Cell Biol 60:523–527

De Vere White R, Tesluk H and Deitcg A (1987) The paradox of aneuploidy in the benign and malignant prostate. Cytometry 1:3

Diamond LW, Nathwani BN and Rappaport H (1982) Flow cytometry in the diagnosis and classification of malignant lymphoma and leukaemia. Cancer 50:1122–1135

Dressler LG and Bartow SA (1989) DNA flow cytometry in solid tumours: practical aspects and clinical applications. Semin Diagn Pathol 6:55–82

Dressler LG, Seamer L, Owens MA et al. (1987) Evaluation of a modelling system for S phase estimation in breast cancer by flow cytometry. Cancer Res 47:5294–5302

Dressler LG, Seamer LC, Owens MA et al. (1988) DNA flow cytometry and prognostic factors in 1331 frozen breast cancer specimens. Cancer 61:420–427

Dressler LG, Eudey L, Gray R et al. (1990) DNA flow cytometry measurements are prognostic for time to recurrence in node negative breast cancer patients: an eastern Co-operative Group (ECOG) intergroup study. Proc Am Soc Clin Oncol 9:22

Durie BGM (1986) Staging and kinetics of multiple myeloma. Semin Oncol 13:300–309

Durie BGM, Salmon SE and Moon TE (1980) Pretreatment tumour mass, cell kinetics and prognosis in multiple myeloma. Blood 55:364–372

Egan MJ, Raafat F, Crocker J and Smith K (1987) Nucleolar organiser regions in small cell tumours of childhood. J Pathol 153:275–280

Egerton DA, Said JM, Epling S and Lee S (1988) DNA content of T-cell lymphomas. A flow cytometric analysis. Am J Pathol 130:326–334

Ellis SJ and Whitehead R (1981) Mitoses counting: a need for reappraisal. Hum Pathol 12:3–4

Ensley J, Maciorowski Z and Pietrazskiewicz H et al. (1990) Prospective correlation of cytotoxic response and DNA content parameters in advanced squamous cell cancer of the head and neck. Proc Am Soc Clin Oncol 9:173

Fallowfield ME and Cook MG (1989) The value of nucleolar organiser region staining in the differential diagnosis of borderline melanocytic lesions. Histopathology 14:229–304

Fallowfield ME, Dodson AR and Cook MG (1988) Nucleolar organiser regions in melanocytic dysplasia and melanoma. Histopathology 13:95–99

Feichter GE, Mueller A and Kaufmann M et al. (1988) Correlations of DNA flow cytometric results and other prognostic factors in primary breast cancer. Int J Cancer 41:823–828

Fisher B, Bauer M, Wickerham L et al. (1983) Relationship of the number of positive axillary nodes to the prognosis of patients with primary breast cancer. Cancer 52:1551–1557

Fossa SD, Thorud E, Shoaib MC et al. (1984) DNA flow cytometry in primary breast carcinoma. Acta Pathol Microbiol Scand 92:475–480

Franzen G, Klintenberg C, Olofsson J and Risberg B (1986) DNA measurement – an objective predictor of response to irradiation. Br J Cancer 53:643

Gentili C, Sanfilippo O and Silvestrini R (1981) Cell proliferation and its relationship to clinical features and relapse in breast cancers. Cancer 48:974–979

Gerdes J, Pickartz H, Brotherton J et al. (1987) Growth fractions and estrogen receptors in human breast cancers as determined in situ with monoclonal antibodies. Am J Pathol 129:486–492

Giri DD, Nottingham JF, Lawry J et al. (1989) Silver binding nucleolar organizer regions (AgNORs) in benign and malignant breast lesions: correlations with ploidy and growth phase by DNA flow cytometry. J Pathol 157:307–313

Grant CS, Hay ID, Dyan JJ et al. (1990) Diagnostic and prognostic utility of flow cytometric DNA measurements in follicular thyroid tumours. World J Surg 14:283–290

Greipp PR and Kyle RA (1983) Clinical, morphological and cell kinetic differences among multiple myeloma, monoclonal gammopathy of undetermined significance and smouldering multiple myeloma. Blood 62:166–171

Haag D, Goerttler M and Tschahargane C (1984) The proliferative index of human breast cancer as obtained by flow cytometry. Pathol Res Pract 178:315–322

Hall PA and Levison DA (1990) Assessment of cell proliferation in histological material. J Clin Pathol 43:184–192

Hall PA, Crocker J, Watts A and Stansfeld AG (1988a) A comparison of nucleolar organizer region staining and Ki-67 immunostaining in non-Hodgkin's lymphoma. Histopathol 12:373–381

Hall PA, Richards MA, Gregory WM et al. (1988b) The prognostic value of Ki-67 immunostaining in non-Hodgkin's lymphoma. J Pathol 154:223–225

Hanby AM, Hall PA, Gregory WM, Dennis P, Rooney N, James P, Richman PI and Levison DA (1990) An inter- and intra-observer study on the diagnosis of lymph node biopsies. J Pathol 161:345a

Hart WR and Billman JK (1978) A reassessment of uterine neoplasms originally diagnosed as leiomyosarcomas. Cancer 41:1902–1910

Hedley DW, Rugg CA and Gelber RD (1987) Association of DNA index and S phase fraction with prognosis in nodes positive breast early breast cancer. Cancer Res 47:4729–4735

Holt S, Croton R, Leinster SJ et al. (1986) In vitro thymidine labelling index in primary operable breast cancer. Eur J Surg Oncol 12:53–57

Kallioniemi O-P, Hietanen T, Mattila J et al. (1987) Aneuploid DNA content and high S phase fraction of tumour cells are related to poor prognosis in patients with primary breast cancer. Eur J Cancer Clin Oncol 23:277–282

Kallioniemi O-P, Punnonen R, Mattila J et al. (1988) Prognostic significance of DNA index, multiploidy and S phase fraction in ovarian cancer. Cancer 61:334–339

Klintenberg C, Stal O, Nordenskjold B et al. (1986) Proliferative index, cytosolic estrogen receptor and axillary node status as prognostic predictors in human mammary cancer. Breast Cancer Res Treat 7(suppl):99–106

Kute TE, Muss HB, Anderson D et al. (1981) Relationship of steroid receptor, cell kinetics and clinical status in patients with breast cancer. Cancer Res 41:3524–3529

Kute TE, Muss HB, Hopkins M et al. (1985) Relationship of flow cytometry results to clinical and steroid receptor status in human breast cancer. Breast Cancer Res Treat 6:113–121

Kvalov S, Morton PF, Kaalhus O et al. (1985) [³H]thymidine uptake in B cell lymphomas. Relationship to treatment response and survival. Scand J Haematol 34:429–435

Lipkin M, Blattner WE, Fraumeni JF et al. (1983) Tritiated thymidine (ψ_p, ψ_h) labelling distribution as a marker for hereditary predisposition to colon cancer. Cancer Res 43:1899–1904

Lykkesfeldt AE, Balsev I, Christensen IJ et al. (1988) DNA ploidy and S phase fraction in primary breast carcinomas in relation to prognostic factors and survival for premenopausal patients at high risk for recurrent disease. Acta Oncol 27:749–756

Macartney JC and Camplejohn RS (1990) DNA flow cytometry in non-Hodgkins lymphomas. Eur J Cancer 26:635–637

Macartney JC, Camplejohn RS, Alder J et al. (1986) Prognostic importance of DNA flow cytometry in non-Hodgkin's lymphomas. J Clin Pathol 39:542–546

Mason BH (1983) Progesterone and estrogen receptors as prognostic variables in breast cancer. Cancer Res 43:2985–2990

Meyer JS and Hixon B (1979) Advanced stage and early relapse of breast carcinomas associated with high thymidine labelling indices. Cancer Res 39:4041–4047

Masters JRW, Camplejohn RS, Millis RR and Rubens RD (1987) Histological grade, elastosis,

DNA ploidy and the response to chemotherapy of breast cancer. Br J Cancer 55:455–457

McDivitt RW, Stone KR, Craig RB and Meyer JS (1985) A comparison of human breast cancer cell kinetics measured by flow cytometry and thymidine labelling. Lab Invest 52:287–291

McDivitt RW, Stone KR, Craig RB et al. (1986) A proposed classification of breast cancer based on kinetic information. Cancer 57:269–276

McFarlane JH, Quirke P and Bird CC (1986) Flow cytometric analysis of DNA heterogeneity in non-Hodgkin's lymphoma. J Pathol 149:236

McGurrin JF, Doria MI, Dawson PJ et al. (1987) Assessment of tumour cell kinetics by immunohistochemistry in carcinoma of breast. Cancer 59:1744–1750

Merkel DE and McGuire WL (1990) Ploidy, proliferative activity and prognosis: DNA flow cytometry of solid tumours. Cancer 65:1194–1295

Meyer JS and Prioleau PG (1981) S phase fractions of colorectal carcinomas related to pathologic and clinical features. Cancer 48:1221–1228

Meyer JS, Friedman E, McCrate MM and Bauer WC (1983) Prediction of early course of breast carcinoma by thymidine labelling. Cancer 51:1879–1886

Meyer JS, Prey MU, Babcock DS and McDivitt RW (1986) Breast carcinome cell kinetics, morphology, stage and host characteristics. Lab Invest 54:41–51

Moran RE, Black MM, Alpert L and Straus MJ (1984) Correlation of cell-cycle kinetics, hormone receptors, histopathology and nodal status in human breast cancer. Cancer 54:1586–1590

Muss HB, Kute TE, Case LD et al. (1989) The relation of flow cytometry to clinical and biologic characteristics in women with node negative primary breast cancer. Cancer 64:1894–1900

NCI Non-Hodgkin's Lymphoma Classification Project Writing Committee (1985) Classification of non-Hodgkin's lymphomas. Cancer 55:9

Nemoto T, Vana J, Bedwani R et al. (1980) Management and survival of female breast cancer: Results of a national survey by the American College of Surgeons. Cancer 45:2917–2924

O'Reilly SM, Camplejohn RS, Barnes DM et al. (1990a) DNA index, S phase fraction, histological grade and prognosis in breast cancer. Br J Cancer 61:671–674

O'Reilly SM, Camplejohn RS, Millis RR et al. (1990b) Node negative breast cancer: prognostic subgroups defined by tumour size and flow cytometry. J Clin Oncol 8:2040–2046

O'Reilly SM, Camplejohn RS, Millis RR et al. (1990c) Proliferative activity, histological grade and benefit from adjuvant chemotherapy in node positive breast cancer. Eur J Cancer 26:1035–1038

Olszewski W, Darzynkiewicz Z, Rosen PP et al. (1981) Flow cytometry of breast carcinoma: relation of tumour cell cycle distribution to histology and estrogen receptor. Cancer 48:985–988

Ploton D, Menager M, Jeannesson P et al. (1986) Improvement in the staining and visualisation of the argyrophilic proteins of the nucleolar organiser region at the optical level. Histochem J 18:5–14

Quinn CM and Wright NA (1990) The clinical assessment of proliferation and growth in human tumours: evaluation of methods and applications as prognostic variables. J Pathol 160:93–102

Raber MN, Barlogie B, Latreille J et al. (1982) Ploidy, proliferative activity and estrogen receptor content in human breast cancer. Cytometry 3:36–41

Rehn S, Glimelius B, Strang P et al. (1990) Prognostic significance of flow cytometry studies in B-cell non-Hodgkin's lymphoma. Haematol Oncol 8:1–12

Remvikos Y, Lavee C, Vilcoq JR et al. (1990) Proliferative activity, response to neo-adjuvant chemotherapy and prognosis in stage II breast cancer. Breast Cancer Res Treat 16:190 (abstr)

Richards MA, Gregory WM and Lister TA (1987) Prognostic factors in lymphoma. In: Stoll BA (ed) Prognostic factors in cancer. Elsevier, Amsterdam, pp 333–357

Schrape S, Jones DB and Wright DH (1987) A comparison of three methods for the determination of the growth fraction in non-Hodgkin's lymphoma. Br J Cancer 55:283–286

Schutte B, Reynders MMJ, Wiggers T et al. (1987) Retrospective analysis of the prognostic significance of DNA content and proliferative activity in large bowel carcinoma. Cancer Res 47:5494–5496

Sciallero S, Bruno S, Di Vinci A et al. (1988) Flow cytometric DNA ploidy in colorectal adenomas and family history of colorectal cancer. Cancer 61:114–120

Sigurdsson H, Baldetorp B, Borg A et al. (1990) Indicators of prognosis in node negative breast cancer. N Engl J Med 322:1045–1053

Silvestrini R, Daidone MG and Gasparini G (1985) Cell kinetics as a prognostic marker in node-negative breast cancer. Cancer 56:1982–1987

Silvestrini R, Costa A, Veneroni S et al. (1988a) Comparative analysis of different approaches to investigate cell kinetics. Cell Tissue Kinet 21:123–131

Silvestrini R, Daidone MG, Selvi et al. (1988b) Cell kinetics as a prognostic indicator in operable breast cancer: updated report. Proc Am Soc Clin Oncol 7:24

Smith R and Crocker J (1988) Evaluation of nucleolar organiser region associated proteins in breast malignancy. Histopathology 12:113–125

Srigley J, Barlogie B, Butler JJ et al. (1985) Heterogeneity of non-Hodgkin's lymphoma probed by nucleic acid cytometry. Blood 65:1090–1096

Stal O, Wingren S, Cartensen J et al. (1989) Prognostic value of DNA ploidy and S phase fraction in relation to estrogen receptor content and clinicopathological variables in primary breast cancer. Eur J Cancer Clin Oncol 25:301–309

Stuart-Harris R, Hedley DW, Taylor IW et al. (1985) Tumour ploidy, response and survival in patients receiving endocrine therapy for advanced breast cancer. Br J Cancer 51:573

Suarez V, Newman J, Hiley C et al. (1989) The value of NOR numbers in neoplastic and non-neoplastic epithelium of the stomach. Histopathology 14:61–66

Sulkes A, Livingston RB and Murphy WK (1979) Tritiated thymidine labelling index and response in human breast cancer. J Natl Cancer Inst 62:513–515

Taylor HB and Norris HJ (1966) Mesenchymal tumours of the uterus. IV. Diagnosis and prognosis of leiomyosarcomas. Arch Pathol 82:40–44

Ten Velde GPM, Schutte B, Roos M et al. (1988) Flow cytometric analysis of DNA ploidy level in paraffin-embedded tissue of non-small cell lung cancer. Eur J Cancer Clin Oncol 24:445–460

Terpstra OT, van Blankenstein M, Dees J and Eilers GU (1987) Abnormal pattern of cell proliferation in the entire colonic mucosa of patients with colon adenoma or cancer. Gastroenterology 92:704–708

Toikkanen S, Joensuu H and Klemi P (1989) The prognostic significance of nuclear DNA content in invasive breast cancer – a study with long-term follow-up. Br J Cancer 60:693–700

Tubiana M and Malaise E (1976) Comparison of cell proliferation kinetics in human and experimental tumours: response to irradiation. Cancer Treat Rep 60:1887–1895

Tubiana M, Pejovic MJ, Renaud A et al. (1981) Kinetic parameters and the course of the disease in breast cancer. Cancer 47:937–943

Straus MJ, Moran R, Muller RE and Wotiz HH (1982) Estrogen receptor heterogeneity and the relationship between estrogen receptor and the tritiated thymidine labelling index in human breast cancer. Oncology 39:197–200

Volm M, Drings P, Mattern J et al. (1985) Prognostic significance of DNA patterns and resistance-predictive tests in non-small cell lung carcinoma. Cancer 56:1396–1403

Walker RA (1988) The histopathological evaluation of nucleolar organiser region proteins. Histopathology 12:221–223

Winter JN, Bauer K, Andersen J et al. (1989) Aneuploidy but not proliferative activity is associated with failure to remit in the diffuse aggressive lymphomas. Proc Am Assoc Cancer Res 30:36

Wooldrige TN, Grierson HL, Weisenburger DD et al. (1988) Association of DNA content and proliferative activity with clinical outcome in patient with diffuse mixed cell and large cell non-Hodgkin's lymphoma. Cancer Res 48:6608–6613

Yang P, Huang GS and Zhu XS (1990) Role of nucleolar organiser regions in differentiating malignant from benign tumours of the colon. J Clin Pathol 43:235–238

11 Cell Proliferation and the Principles of Cancer Therapy

W.M. GREGORY

Evolution of the Principles

The guiding principles on which a great deal of modern cancer therapy is based originate with experiments undertaken by Skipper and colleagues in the 1960s (Skipper et al. 1964, 1967). As a result of his investigations of experimental tumour systems, Skipper concluded that a given dose or course of (chemo)therapy will kill a constant *fraction* of the cell population, rather than a constant number of cells. He went on to consider the consequences of tumours having different growth rates, and concluded that chemotherapy gave a greater fractional cell-kill to more rapidly growing tumours, and was more likely to be curative in such cases (Skipper and Perry 1970). It also appeared that larger tumours generally grew more slowly, and that the lack of responsiveness of such tumours was related to this slower rate of proliferation (Shackney 1970; Steel et al. 1976).

These results apply to radiotherapy as well as chemotherapy (see for instance, Okumura et al. 1977), and have now been explored in considerable detail. Larger tumours generally grow more slowly because although early (subclinical) tumour growth follows an exponential pattern, tumour growth slows down as the tumour increases in size, presumably due to problems of nutrient supply, approaching a maximum volume. Laird (1964) proposed a gompertzian function to describe this growth, where the (exponential) growth rate also declines exponentially, resulting in such a maximum volume. The gompertzian model appears to fit the experimental data well for a variety of tumours and growth rates (for example: Akanuma 1978; Demicheli 1980; Pearlman 1983; Sullivan and Salmon 1972). Thus the unresponsiveness of a tumour may be highly dependent on the point in its growth curve at which therapy is initiated.

It is interesting to note that with a gompertzian growth curve, effective treatment of large indolent tumours can result in smaller, more rapidly dividing tumours, with new possibilities for treatment. Norton and Simon (1977, 1986) suggested that initial induction treatments, at modest doses, could be used merely as a method of reducing the tumour to a size where it grew more rapidly. At this point, intensive therapy could be initiated in an attempt at cure.

Following on from these early principles, Goldie and Coldman (1979) examined the success or failure of therapy from a different point of view, namely the presence or the acquisition of resistance. They demonstrated that if there was a constant rate of mutation towards resistance (i.e. if every time a cell divided it had the same chance of mutating to become resistant), then there would be a critical and *short* period in the tumour's history when the chances of cure dropped from one to zero. (Essentially there would be a critical "mass" of tumour cells beyond which the chances of a mutation occurring would be very high.) Examples were given where this occurred in a 1-log range, for example from 10^7 cells to 10^8 cells. Thus it would be vital to treat during, or preferably before, this period. Of course, the more rapidly the tumour grew, the more important it would be to treat early, as the tumour would be quickly progressing to the point where resistant mutants would inevitably arise.

To summarise these results, the intensity, frequency and duration of therapy should be matched to the tumour's growth rate and to the point reached in the tumour's growth curve (essentially the tumour size). The more rapid the growth rate, the more intensive, frequent and short lived should be the therapy. For slowly growing tumours, therapy of longer duration will be necessary, probably at reduced doses. The aim, in these latter cases, should be to eliminate the dividing cells, which may be a small fraction of the tumour. Higher doses are likely to be unproductive, killing few extra cells and possibly compromising later therapy due to toxicity, acquisition of resistance, etc. Additionally, therapy should be given as early as possible, to maximise the chances of cure, by treating before resistant mutants have arisen. This will be especially critical in rapidly growing tumours, where a short delay could allow a large increase in tumour size, and consequently a severe reduction in the chances of cure.

The Principles in Practice

The principles outlined above can be applied not only to the choice of treatments for different cancers but also, on the basis of growth rates of their tumours, to the choice of treatment for different individuals with the same cancer. A number of examples will be given to demonstrate these points.

Application to Particular Neoplasms

Consider firstly, a very rapidly growing malignancy such as testicular teratoma; here the therapist should aim to administer the maximum dose in the minimum time. Cure rates for patients with stage-IV metastatic disease have increased dramatically, rising from less than 10% in the 1960s, when treatment was spread over 2 or more years (Mackenzie et al. 1966; McElwain et al. 1974) to 80%–90% in the late 1970s and 1980s with the introduction of high doses of new and more effective platinum-based regimens, typically administered for only two or three courses (Newlands et al. 1983; Oliver 1986; Peckham et al. 1983; Vugrin et al. 1983).

In contrast, consider the treatment of gastric cancer. Initial adjuvant studies comprised moderate doses of mitomycin C spread over a period of 5 weeks, and achieved modest but definite success (Imanaga and Nakazato 1977; Nakajima et al. 1978). With the popularity of intensive scheduling, subsequent trials administered larger doses over shorter time periods, including treatment concurrent with surgery (Imanaga and Nakazato 1977). These trials not only failed to improve on the earlier results, but any beneficial effect appeared to have been lost. Examination of times taken for early untreated gastric cancer patients to progress to late stage (Tsukuma et al. 1983) demonstrate that gastric cancer has a slow growth rate, and thus longer and sustained therapy is necessary. It may be that 5 weeks is inadequate, and that further benefit would be gained by even more prolonged therapy.

It is interesting to note that, for many cancers, trials addressing the question of duration of treatment are now being undertaken. For instance, in Wilms' tumour, as a result of a series of randomised trials over the last 21 years, the standard duration of therapy has been reduced from 15 months to 10 weeks (D'Angio et al. 1976, 1981, 1989). Similar trials and comparisons have also been undertaken in Hodgkin's disease (De Vita et al. 1980; Medical Research Council's Working Party on Lymphomas 1979; Young et al. 1973), non-Hodgkin's lymphomas (Connors and Klimo 1988) and leukaemia (Bell et al. 1982; Vaughan et al. 1984) among others, supporting the use of short intensive induction regimes without maintenance therapy in these diseases.

In early trials of adjuvant therapy for patients with breast cancer, the typical duration of treatment was 12 months (Bonadonna et al. 1985; Bonadonna and Valagussa 1987; Richards et al. 1990). It has subsequently been shown that equivalent results can be achieved with only 6 months treatment. However, trials in which prolonged (6 months or more) treatment was compared with single-course (perioperative) treatment have demonstrated that prolonged treatment is more effective (unpublished results presented at the Early Breast Cancer Trialists' Collaborative Group meeting, Oxford, 1990). It is still unclear whether, for example, 3 months treatment would be as effective as 6 months treatment. Again it is apparent that breast cancer is a relatively slowly growing tumour. For instance, using the monoclonal antibody Ki-67, which is reported to stain cells not in the G_0 phase of the cell cycle and to give reliable estimates for the number of proliferating cells in tumours, less than 20% of breast cancer cells are stained on average (Barnard et al. 1987; Gerdes et al. 1986). This compares with an average of greater than 50% in, for example, high-grade non-Hodgkin's lymphomas (Gerdes et al. 1984). Once again, moderate-dose treatment of longer duration seems to be required.

Minimising the duration of therapy may, on the surface, appear relatively unimportant, and indeed dangerous, since some of the efficacy may be lost. Once a successful therapy has been introduced, and perhaps shown to be effective in a randomised trial, it often survives largely unaltered for many years, since clinicians fear that tampering may abrogate the effect. The administration of long-term maintenance therapy in acute lymphoblastic leukaemia is one example among many (Freireich et al. 1963); this is covered later in more detail. This conservatism, particularly with respect to treatment duration, should be resisted. In the laboratory, long-term low-dose therapy is a classical method of developing resistant cell lines. It also seems likely that the effectiveness of relapse therapy will be impaired by longer initial durations of treatment. Furthermore, the additional toxicity produced by further therapy may not merely be undesirable, it may also have implications for the patients' psychological outlook, their immune response, and thus their chances of relapse (Greer and Watson 1987; Ramirez et al. 1989).

Application in Choice of Treatment for Individuals

There are two distinct approaches to application of the above principles to the choice of treatment for individuals. One is to use growth rate related prognostic factors to delineate ever smaller groups of patients with different growth rates. For instance, by choosing patients based on grade of tumour or S-phase fraction measurements in breast cancer, it is possible to target a group of early-stage high-risk patients with aggressive tumours, for whom intensive therapy may be appropriate. (This is covered in detail in Chap. 10 by O'Reilly and Richards.) Of course, within these subgroups, there will still be a wide range of different growth rates (see for example, the distribution of Ki-67 values for different histological subtypes of non-Hodgkin's lymphoma, lung and breast cancer; Barnard et al. 1987; Gatter et al. 1986; Gerdes et al. 1984). If possible, a better approach would be to estimate the growth rates of individual tumours before treatment, to choose dose and duration appropriately, and ideally to monitor response, in order to determine when treatment was no longer effective or necessary.

Diseases where choice of treatment based on proliferation values of individual tumours would be especially appropriate and effective, are those with a wide range of growth rates, and an average growth rate that is relatively high. One such malignancy is non-Hodgkin's lymphoma. Prospective studies are unfortunately lacking, but analysis of response and duration of response in patients studied with the Ki-67 antibody and labelling index techniques suggests that, if treated early and intensively, response rates are higher in the more rapidly proliferating tumours, and durations of response are different (Hall et al. 1988). Relapses in rapidly growing tumours occur early, and those surviving this phase tend to be cured (or remain disease-free much longer). Relapses in slowly growing tumours may occur late and be spread over a long time interval.

Various models have been proposed to monitor tumour response by means of repeated tumour volume estimates (Birkhead and Gregory 1984; Birkhead et al. 1987) or tumour marker levels (Price et al. 1990a, b) and to infer resistance and growth rate parameters during treatment. Applications have suggested that such models could be used to determine when to stop treatment for an individual, following elimination of all but resistant disease, and possibly switch to an (hopefully non-cross-resistant) alternative (Gregory et al. 1988, 1990).

New Approaches to Estimating Resistance and Growth Rates

Although the principles outlined above have been useful in designing treatment regimens and strategies, their application has often proved slow and laborious. It may take numerous trials to establish the optimum number of courses of treatment to administer, as demonstrated by trials in Wilms' tumour for instance (D'Angio et al. 1976, 1981, 1989), or to establish what doses are required to achieve optimum cell-kill, quite apart from the problem of which drugs to combine in the first place. This is partly because clinical trials are designed merely to discover whether one treatment is better than another, and not why or to what degree it is better. Nevertheless, there has been a gradual development of mathematical models that attempt to provide this important information. These have now advanced to the point where this vital information regarding why and to what extent one treatment is superior to another is obtainable, and this information is proving extremely instructive. A number of initial applications have already proved fruitful, and will be examined in some detail.

Mathematical Model Description and Development

Shackney (1978) reviewed the growth rates of a wide range of cancers, and examined the effects on relapse patterns. He showed that rapidly growing tumours had response-duration curves that fell steeply, and that, conversely, slowly growing tumours showed patterns of gradual relapse over long periods. Measurements of growth rates showed log-normal distributions of doubling times within particular cancers, and these were also related to the relapse pattern. Norton (1988) developed a mathematical model to show how different sized tumours with these growth patterns would produce systematically different response-duration curves.

Although these models aided in the explanation of trial results, they were still essentially subjective in nature, inviting interpretation from the shapes of response duration curves. A recent mathematical model (Gregory et al. 1991), enables the relapse times themselves to be used to estimate the volume and growth rates of resistant disease, thus bringing in objectivity and enabling more widespread application. This model will now be described in more detail and examples given to show its use in understanding trial results, designing new treatments, and reinterpreting the results of previous studies.

The model assumes that both the distribution of resistant disease and the distribution of growth rates are log-normally distributed, based respectively on the Goldie–Coldman mutation-to-resistant hypothesis (Goldie and Coldman 1979) and the examination of tumour growth rates for groups of patients with the same cancer (Gerdes et al. 1984, 1986; Shackney 1978). By deriving a mathematical formula for the response-duration curve based on these assumptions, the model can "fit" any given curve (see, for example, Fig. 11.1), producing best estimates for the mean number of resistant cells, the spread (standard deviation) in the number of resistant cells, the mean growth rate, and the spread of growth rates. Confidence limits can also be produced for these estimates. The results can be shown in graphical form, by plotting the distribution of the log of resistant tumour, as in Fig. 11.2. (The technical details can be found in Gregory et al. 1991.)

The model thus appears to have the potential for evaluating the cell-kills of treatments in vivo. In addition, because application of the model requires actuarial response-duration data only, it is possible to re-evaluate old clinical trials, gaining additional information on tumour volume and growth rate.

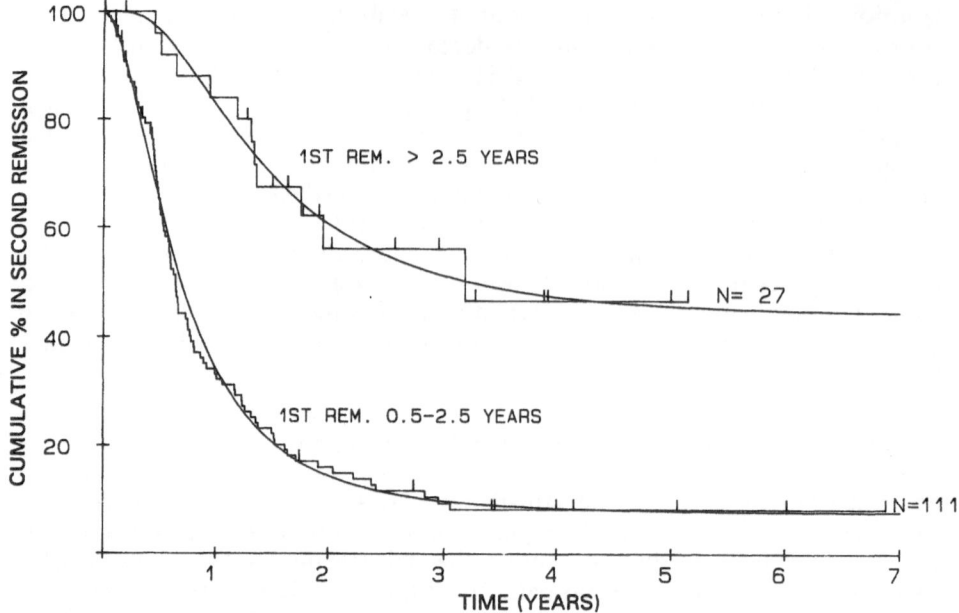

Figure 11.1. The Medical Research Council acute myelogenous leukaemia trial (Rees et al. 1986): durations of second remission (stepped curve) as they relate to the duration of first remission, with model fits (smooth curve) (Gregory et al. 1991).

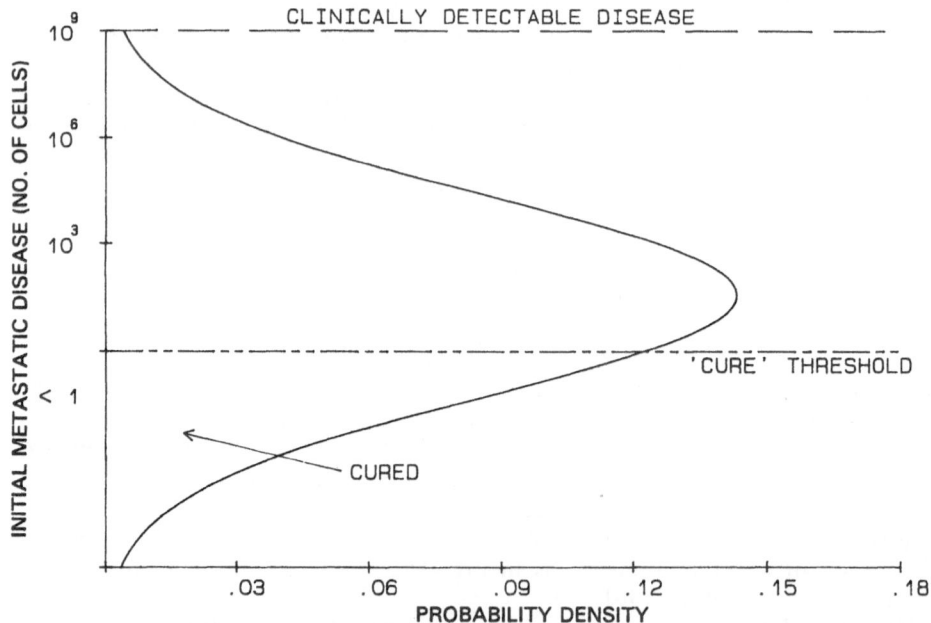

Figure 11.2. EORTC locally advanced breast cancer trial (Rubens et al. 1989): estimated distribution of resistant subclinical metastatic tumour at the time of initial diagnosis (Gregory et al. 1991).

Model Applications

Acute Myelogenous Leukaemia

In an attempt to validate the model, information obtained from model application to durations of second complete remission (CR) was compared with that obtained from first CR in acute myelogenous leukaemia. A Medical Research Council trial of maintenance therapy after induction and consolidation with daunorubicin, cytosine arabinoside, and 6-thioguanine (DAT) was examined (Rees et al. 1986). The trial was sufficiently large (1127 patients) to have enough second remitters for detailed analysis. From the first remission times, the model estimated that patients relapsing beyond 2.5 years should have had more slowly growing disease than the rest, with a very small volume of resistant disease (see Gregory et al. 1991). On examination, the durations of second CR (shown in Fig. 11.1, along with the model fits) were indeed related to the durations of first CR, with long first remitters having longer durations of second CR, and a better response to second-line therapy. Furthermore, on application to the durations of second CR, the model estimated that these patients had more slowly growing tumours than the rest (doubling time 19 days compared with 14) and a very small volume of resistant disease. The results were thus consistent. Furthermore, it could be surmised that patients with long first remissions could be appropriately treated with more curative intent on relapse, perhaps with a slightly longer duration of more intensive therapy than that currently employed.

Breast Cancer

The model has been applied to data from an EORTC study (Rubens et al. 1989) of treatment for locally advanced breast cancer. The effect of chemotherapy and endocrine therapy on cell-kill and growth rates of both locoregional and distant metastatic disease was examined. Chemotherapy appeared to give a 2-log cell-kill in the local tumour, but had little effect on distant metastatic disease. Endocrine therapy, in contrast, appeared to arrest growth for a period, followed by continued regrowth and relapse, in addition to having a slight cell-killing effect. This clearly has important consequences for combined therapy in these patients. If both chemotherapy, which kills mainly dividing cells, and hormone therapy, which results in growth arrest, are to be administered, the chemotherapy must be given first, otherwise the chemotherapy will have few dividing cells to kill, and is likely to prove ineffective.

When applied to the times to occurrence of distant metastatic disease, the mean volume of resistant subclinical metastatic tumour present at the time of initial diagnosis was estimated. The results suggested that this mean volume of subclinical metastatic disease was smaller than expected – approximately 2 logs, as shown in Fig. 11.2. This suggests a new treatment strategy for these patients. As standard chemotherapy (CMF) appears to be ineffective in these

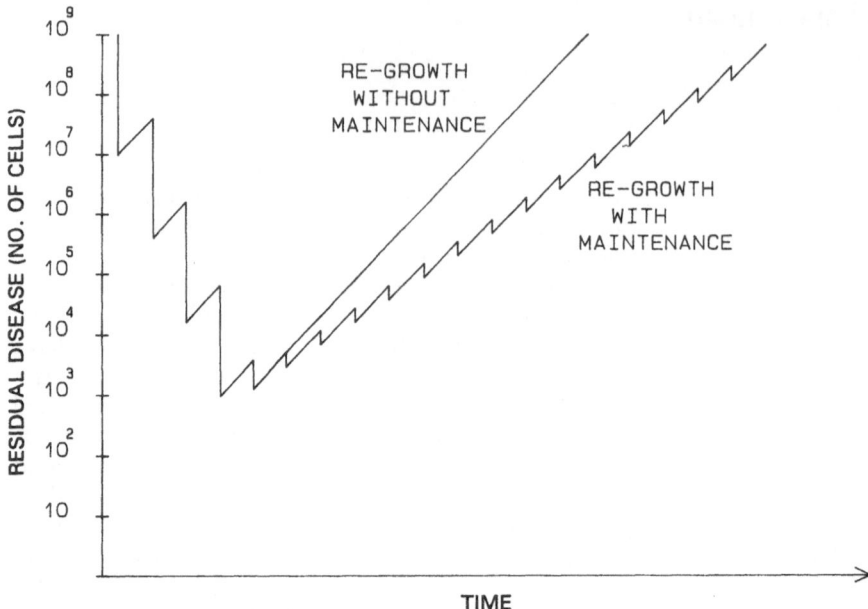

Figure 11.3. Hypothesised effect of maintenance therapy on cell-kill during tumour re-growth.

circumstances, alternative strategies are required. Possibilities include the use of high-dose chemotherapy or total body irradiation supported by bone marrow transplant. Recruitment of cells into the cell cycle by stimulation with oestrogens or other factors may also render them more sensitive to subsequent cytotoxic therapy.

Maintenance Therapy

The model has also been applied to an early trial (Freireich et al. 1963) of 6-mercaptopurine (6-MP) maintenance therapy in acute lymphoblastic leukaemia (ALL). It was originally concluded from this trial that 6-MP maintenance therapy delayed relapse in ALL. This was presumed to be a result of a slow-down or arrest in the tumour's re-growth. This form of long-term, low-dose treatment has been employed almost universally in ALL ever since. However, results from the model suggested that 6-MP actually worked by achieving a nearly 4-log cell-kill, and that the delay before relapse was a result of the time taken for this tumour to re-grow to clinically detectable levels (see Gregory et al. 1991). It appears from this analysis that 6-MP should be used as a remission induction, rather than remission maintenance agent.

This result for the early ALL study casts doubt on the whole concept of maintenance therapy, since it was from this sort of study that the principle was derived. This example gives an unambiguous result since there are no confounding factors, such as other drugs, radiation therapy, etc. However, maintenance therapy is frequently given after administration of multi-drug intensive therapy when considerable cell-kill has already been accomplished (for example, see Rees et al. 1986), and its effect is thus more difficult to determine.

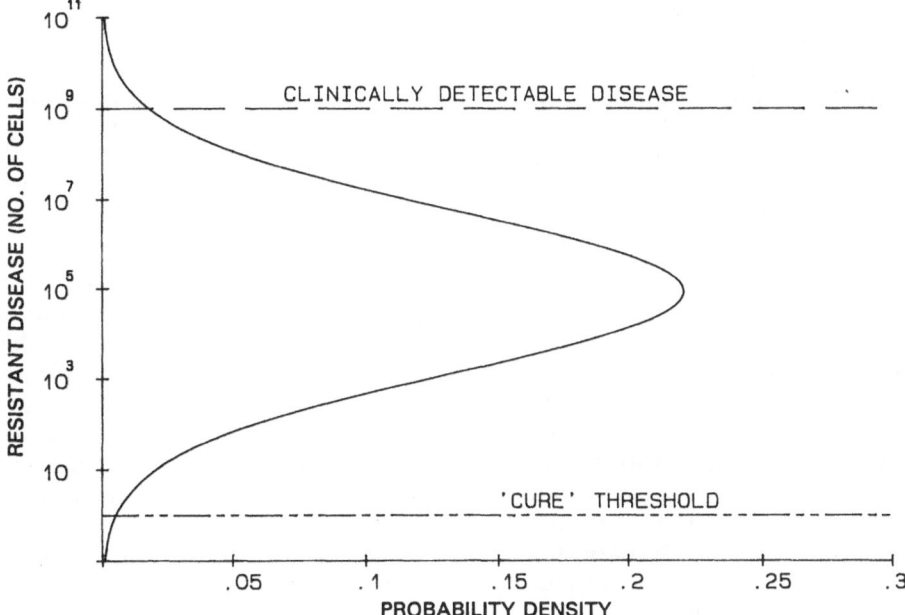

Figure 11.4. Estimated distribution (Gregory et al. 1991) of resistant disease in multiple myeloma patients treated with high-dose melphalan (Selby et al. 1987).

If the 6-MP therapy is really maintaining remission, by delaying or slowing down re-growth, how is this being achieved? A likely explanation is that the treatment has a continued slight cell-killing effect on the remaining tumour, so that it appears as if the re-growth rate has been slowed (see Fig. 11.3). If this hypothesis is correct, and the drugs are actually killing tumour cells over many courses, then the treatment ought to have curative potential. However, maintenance trials never seem to increase the proportion of patients cured. It is suggested that this is because the doses given kill fewer cells than are replenished by tumour growth between courses (as in Fig. 11.3). Thus the doses given will be inadequate to affect the chances of cure.

It is hard to resist the inference that the logical role for an apparently effective maintenance agent should instead be as an induction or consolidation agent, presumably given in greater doses, over a much shorter period of time.

Multiple Myeloma

Another application of the model concerns administration of very intensive chemotherapy in multiple myeloma. Complete responses were rarely seen in this disease until the recent administration of high-dose melphalan. The model was applied to the complete response-duration curves for 41 patients treated with high-dose mephalan at St. Bartholomew's and the Royal Marsden hospitals (Selby et al. 1987). The estimated picture of residual disease is shown in Fig. 11.4. The model estimated that, although high dose melphalan achieves a mean cell-kill of about 4 logs, a further 5 logs would be necessary, on average, to cure these patients. Thus minor alterations to the regime, such as further slight increases in melphalan dose would be unlikely to achieve cure. Additional, more drastic alterations appeared to be necessary. Subsequent regimens have therefore employed combination chemotherapy for a median of five courses before (late) administration of high-dose melphalan (Gore et al. 1989). Early results from these trials appear encouraging, although longer follow-up is required to discover whether cures have been achieved.

Hodgkin's Disease

The model was applied to 89 patients with stage-I Hodgkin's disease treated at St. Bartholomew's hospital (Ganesan et al. 1990). All patients were treated with radiotherapy alone, and the 15-year disease-free rate, taken from the actuarial curve, was 74%. It was observed that patients destined to relapse probably had only 2 logs or less of malignant cells remaining after treatment. Since six courses of combination chemotherapy (e.g. MVPP) cures the majority of later stage patients with Hodgkin's disease, it appears likely that employing a single course of such chemotherapy would cure virtually all patients with stage-I disease. Currently chemotherapy is rarely used in stage-I Hodgkin's disease because of the resulting toxicity, particularly the risk of sterility (Waxman et al. 1987). One course would be likely to cause very little toxicity, for a seemingly significant benefit.

Conclusion

Clinicians have been struggling for some 45 years with the problems of how best to use the (chemo)therapeutic agents at their disposal. Many drugs are dramatically effective at killing tumour cells and inducing complete disease remissions. However, relapse is still the norm in many cancers. This is a very frustrating picture, since it is often felt that relatively minor changes in dose and/or scheduling might produce striking improvements. Indeed, even recently, alterations in dose and scheduling of drugs which have been in use for many years have shown great effects (Slevin et al. 1989). The key to these improvements appears to lie in the exploitation of the growth kinetics of tumours, by timing the administration to hit cycling cells, and matching dose and schedule to different cancers, and ultimately to the individual patient. The principles outlined for the use of cancer therapy by Skipper and colleagues, and since built upon by many others, would enable a more appropriate choice of treatment if more information was available on tumour growth rates, acquisition of resistance and cell-kill.

With the advent of multi-drug chemotherapy, the above situation was further exacerbated. The choice of drug combinations and schedules is often bewilderingly large; consider for example some of the trials employed for Hodgkin's and non-Hodgkin's lymphomas (Bonadonna et al. 1986; Fisher et al. 1983; Klimo et al. 1985; Skarin et al. 1983). With clinical trials only addressing the simple question of whether one regimen is better than another (and this often ineffectively), progress is slow, and the literature confusing, with many puzzling and apparently contradictory results (see, for example, Slevin et al. 1986).

One remedy to this situation lies in the use of recently developed mathematical models, which attempt to provide the kind of information from a trial which leads to an understanding of *why* differences occurred. This information may help in the design of future trials, and directions for research. Thus, trials should be able to progress in a more structured and rational manner. Combining these models with recently developed methods for accurately assessing tumour proliferation (described elsewhere in this book) should further enhance the development of effective treatments. It is to be hoped that these models will be further applied and developed to optimise the use of current treatments, as well as those now under development.

References

Akanuma A (1978) Parameter analysis of gompertzian function growth model in clinical tumours. Eur J Cancer 14:681–688

Barnard NJ, Hall PA, Lemoine NJ and Kadar N (1987) Proliferative index in breast carcinoma determined in situ by Ki67 immunostaining and its relationship to clinical and pathological variables. J Pathol 152:287–295

Bell R, Rohatiner AZS, Slevin ML et al. (1982) Short-term treatment for acute myelogenous leukaemia. Br Med J 284:1221–1229

Birkhead BG and Gregory WM (1984) A mathematical model of the effects of drug resistance in cancer chemotherapy. Math Biosci 72:59–70

Birkhead BG, Rankin EM, Gallivan S, Dones L and Rubens RD (1987) A mathematical model of the development of drug resistance to cancer chemotherapy. Eur J Cancer Clin Oncol 23: 1421–1427

Bonadonna G, Valagussa P, Rossi A et al. (1985) Ten-year experience with CMF-based adjuvant chemotherapy in resectable breast cancer. Breast Cancer Res Treat 5:95–115

Bonadonna G and Valagussa P (1987) Current status of adjuvant chemotherapy for breast cancer. Semin Oncol 14:8–22

Bonadonna G, Valagussa P and Santoro A (1986) Alternating non-cross resistant combination chemotherapy or MOPP in stage IV Hodgkin's disease. Ann Intern Med 104:739–746

Connors JM and Klimo P (1988) MACOP-B chemotherapy for malignant lymphomas and related conditions: 1987 update and additional observations. Semin Haematol 25:41–46

D'Angio GJ, Evans AE, Breslow N et al. (1976) The treatment of Wilms' tumour. Results of the National Wilms' Tumour Study. Cancer 38:633–646

D'Angio GJ, Evans A, Breslow N et al. (1981) The treatment of Wilms' tumour. Results of the Second National Wilms' Tumour Study. Cancer 47:2302–2311

D'Angio GJ, Breslow N, Beckwith JB et al. (1989) Treatment of Wilms' tumor. Results of the Third National Wilms' Tumor Study. Cancer 64:349–360

Demicheli R (1980) Growth of testicular neoplasm lung metastases: tumour specific correlation between two Gompertzian parameters. Eur J Cancer 16:1603–1608

De Vita VT, Simon RM, Hubbard SM et al. (1980) Curability of advanced Hodgkin's disease with chemotherapy. Long term follow-up of MOPP-treated patients at the National Cancer Institute. Ann Intern Med 92:587–594

Fisher RI, De Vita VT Jr and Hubbard SM (1983) Diffuse aggresive lymphomas: increased survival after alternating flexible sequences of ProMACE and MOPP chemotherapy. Ann Intern Med 98:304–309

Freireich EJ, Gehan E, Frei III E et al. (1963) The effect of 6-mercaptopurine on the duration of steroid-induced remissions in acute leukaemia: a model for evaluation of other potentially useful therapy. Blood 21:699–716

Ganesan TS, Wrigley PF, Murray PA et al. (1990) Radiotherapy for stage I Hodgkin's disease: 20 years experience at St Bartholomew's Hospital. Br J Cancer 62:314–318

Gatter KC, Dunnill MS, Gerdes J, Stein H and Mason DY (1986) New approach to assessing lung tumours in man. J Clin Pathol 39:590–593

Gerdes J, Dallenbach F, Lennert K, Lemke H and Stein H (1984) Growth fractions in malignant non-Hodgkin's lymphoma as determined in situ by the monoclonal antibody Ki67. Haematol Oncol 2:365–371

Gerdes J, Lelle RJ, Rickartz H et al. (1986) Growth fractions in breast cancers determined in situ with monoclonal antibody Ki67. J Clin Pathol 39:977–980

Goldie JH and Coldman AJ (1979) A mathematical model for relating the drug sensitivity of tumours to their spontaneous mutation rate. Cancer Treat Rep 63:1727–1731

Gore ME, Selby PJ, Viner C et al. (1989) Intensive treatment of multiple myeloma and criteria for complete remission. Lancet ii:879–882

Greer S and Watson M (1987) Mental adjustment to cancer: its measurement and prognostic importance. Cancer Surv 6:439–453

Gregory WM, Birkhead BG and Souhami RL (1988) A mathematical model of drug resistance applied to treatment for small cell lung cancer. J Clin Oncol 6:457–461

Gregory WM, Reznek RH, Hallett M and Slevin ML (1990) Using mathematical models fo estimate drug resistance and treatment efficacy via CT scan measurements of tumour volume. Br J Cancer 62:671–675

Gregory WM, Richards MA, Slevin ML and Souhami RL (1991) A mathematical model relating response durations to amount of sub-clinical resistant disease. Cancer Res 51:1210–1216

Hall PA, Richards MA, Gregory WM, d'Ardenne AJ, Lister TA and Stansfeld AG (1988) The prognostic value of Ki67 immunostaining in Non-Hodgkin's lymphoma. J Pathol 154:223–236

Imanaga H and Nakazato H (1977) Results of surgery for gastric cancer and effect of adjuvant mitomycin C on cancer recurrence. World J Surg 1:213–221

Klimo P and Connors JM (1985) MOPP/ABV hybrid program: combination chemotherapy based on early introduction of seven effective drugs for advanced Hodgkin's disease. J Clin Oncol 3:1174–1182

Laird AK (1964) Dynamics of tumour growth. Br J Cancer 18:490–502

Mackenzie AR (1966) Chemotherapy of metastatic testis cancer: results in 54 patients. Cancer 19:1369–1376

McElwain TJ and Peckham MJ (1974) Combination chemotherapy in testicular tumours. Proc R Soc Med 67:297

Medical Research Council's Working Party on Lymphomas (1979) Randomised trial of two-drug and four-drug maintenance chemotherapy in advanced or recurrent Hodgkin's disease. Br Med J 1:1105–1108

Nakajima T, Fukami A, Ohashi T and Kajitani T (1978) Long-term follow-up study of gastric cancer patients treated with surgery and adjuvant chemotherapy with mitomycin C. Int J Clin Pharmacol Biopharm 16:209–216

Newlands ES, Begent RHJ, Rustin GJS, Parker D and Bagshawe KD (1983) Further advances in the management of malignant teratomas of the testis and other sites. Lancet i:948

Norton L (1988) A gompertzian model of human breast cancer growth. Cancer Res 48:7067–7071

Norton L and Simon R (1977) Tumour size, sensitivity to therapy, and design of treatment schedules. Cancer Treat Rep 61:1307–1317

Norton L and Simon R (1986) The Norton–Simon hypothesis revisited. Cancer Treat Rep 70: 163–169

Oliver RTD (1986) Germ cell tumours. In: Slevin ML and Staquet MJ (eds) Randomised trials in cancer: a critical review by sites. Raven Press, New York

Okumura Y, Ueda T, Mori T and Kitabatake T (1977) Kinetic analysis of tumour regression during the course of radiotherapy. Strahlentherapie 153:35–39

Pearlman AW (1983) Doubling time and survival time. In: Stoll BA (ed) Cancer treatment: endpoint evaluation. John Wiley and Sons, New York, pp 279–301.

Peckham MJ, Barrett A and Lieu KH (1983) The treatment of metastatic germ cell testicular tumours with bleomycin, etoposide and cisplatinum (BEP). Br J Cancer 47:613–619

Price P, Hogan SJ and Horwich A (1990a) The growth rate of metastatic non-seminomatous germ cell testicular tumours measured by marker production doubling time – I. Theoretical basis and practical application. Eur J Cancer 26:450–453

Price P, Hogan SJ, Bliss JM and Horwich A (1990b) The growth rate of metastatic nonseminomatous germ cell testicular tumours measured by marker production doubling time – II. Prognostic significance in patients treated by chemotherapy. Eur J Cancer 26:453–457

Ramirez AJ, Craig TK, Watson JP, Fentiman IS, North WR and Rubens RD (1989) Stress and relapse of breast cancer. Br Med J 298:291–293

Rees JKH, Gray RG, Swirzky D and Hayhoe FGJ (1986) Principal results of the medical research council's 8th acute myeloid leukaemia trial. Lancet ii:1236–1241

Richards MA, O'Reilly SM, Howell A et al. (1990) Adjuvant cyclophosphamide, methotrexate, and fluorouracil in patients with axillary node-positive breast cancer: an update of the Guy's/Manchester trial. J Clin Oncol 8:2032–2039

Rubens RD, Bartelink H and Engelsman E et al. (1989) Locally advanced breast cancer: the contribution of cytotoxic and endocrine treatment to radiotherapy. Eur J Cancer Clin Oncol 25:667–678

Selby PJ, McElwain TJ, Nandi AC et al. (1987) Multiple myeloma treated with high dose intravenous melphalan. Br J Haematol 66:55–62

Shackney SE (1970) A computer model for tumour growth and chemotherapy, and its application to L1210 leukemia treated with cytosine arabinoside (NSC-63878). Cancer Chemother Rep 54:399–429

Shackney SE, McCormack GW and Cuchural GJ (1978) Growth patterns of solid tumours and their relation to responsiveness to therapy. Ann Intern Med 89:107–121

Skarin AT, Canellos GP and Rosenthal DS (1983) Improved prognosis of diffuse histiocytic and undifferentiated lymphoma by use of high dose methotrexate alternating with standard agents (M-BACOD). J Clin Oncol 1:91–98

Skipper HE and Perry S (1970) Kinetics of normal and leukemic leukocyte populations and relevance to chemotherapy. Cancer Res 30:1883–1897

Skipper HE, Schabel FM Jr and Wilcox WS (1964) Experimental evaluation of potential anticancer agents. XIII. On the criteria and kinetics associated with "curability" of experimental leukemia. Cancer Chemother Rep 35:1–111

Skipper HE, Schabel FM Jr and Wilcox WS (1967) Experimental evaluation of anticancer agents. XXI. Scheduling of arabinosylcytosine to take advantage of its S-phase specificity against leukemia cells. Cancer Chemother Rep 51:125–165

Slevin ML and Staquet MJ (eds) (1986) Randomised trials in cancer: a critical review by sites. Raven Press, New York

Slevin ML, Clark PI, Joel SP et al. (1989) A randomised trial to evaluate the effect of schedule on the activity of etoposide in small cell lung cancer. J Clin Oncol 7:1333–1340

Steel GG, Adams K and Stanley J (1976) Size-dependence of the response of Lewis lung tumours to BCNU. Cancer Treatment Reports 60:1743–1748

Sullivan PW and Salmon SE (1972) Kinetics of tumour growth and regression in IgG multiple myeloma. J Clin Invest 51:1967–1973

Tsukuma H, Mishima T and Oshima A (1983) Prospective study of "early" gastric cancer. Int J Cancer 31:421–426

Vaughan WP, Karp JE and Burke PJ (1984) Two-cycle timed sequential chemotherapy for adult acute nonlymphocytic leukemia. Blood 64:975–980

Vugrin D, Whitmore WE Jr and Golbey RB (1983) VAB-6 combination of chemotherapy without maintenance in treatment of disseminated cancer of the testis. Cancer 51:211

Waxman JH, Ahmed R, Smith D et al. (1987) Failure to preserve fertility in patients with Hodgkin's disease. Cancer Chemother Pharmacol 19:159–162

Young RC, Canellos GP, Chabner BA, Schein PS and De Vita VT (1973) Maintenance chemotherapy for advanced Hodgkin's disease in remission. Lancet i:1339–1343

Subject Index